# U.S.NRC

United States Nuclear Regulatory Commission

*Protecting People and the Environment*

NUREG/CR-7144
SAND 2008 - 3938

I0494090

# Laminar Hydraulic Analysis of a Commercial Pressurized Water Reactor Fuel Assembly

Office of Nuclear Regulatory Research

## AVAILABILITY OF REFERENCE MATERIALS
## IN NRC PUBLICATIONS

### NRC Reference Material

As of November 1999, you may electronically access NUREG-series publications and other NRC records at NRC's Public Electronic Reading Room at http://www.nrc.gov/reading-rm.html. Publicly released records include, to name a few, NUREG-series publications; *Federal Register* notices; applicant, licensee, and vendor documents and correspondence; NRC correspondence and internal memoranda; bulletins and information notices; inspection and investigative reports; licensee event reports; and Commission papers and their attachments.

NRC publications in the NUREG series, NRC regulations, and Title 10, "Energy," in the *Code of Federal Regulations* may also be purchased from one of these two sources.
1. The Superintendent of Documents
   U.S. Government Printing Office Mail Stop SSOP
   Washington, DC 20402–0001
   Internet: bookstore.gpo.gov
   Telephone: 202-512-1800
   Fax: 202-512-2250
2. The National Technical Information Service
   Springfield, VA 22161–0002
   www.ntis.gov
   1–800–553–6847 or, locally, 703–605–6000

A single copy of each NRC draft report for comment is available free, to the extent of supply, upon written request as follows:
Address: U.S. Nuclear Regulatory Commission
         Office of Administration
         Publications Branch
         Washington, DC 20555-0001
E-mail: DISTRIBUTION.RESOURCE@NRC.GOV
Facsimile: 301–415–2289

Some publications in the NUREG series that are posted at NRC's Web site address http://www.nrc.gov/reading-rm/doc-collections/nuregs are updated periodically and may differ from the last printed version. Although references to material found on a Web site bear the date the material was accessed, the material available on the date cited may subsequently be removed from the site.

### Non-NRC Reference Material

Documents available from public and special technical libraries include all open literature items, such as books, journal articles, transactions, *Federal Register* notices, Federal and State legislation, and congressional reports. Such documents as theses, dissertations, foreign reports and translations, and non-NRC conference proceedings may be purchased from their sponsoring organization.

Copies of industry codes and standards used in a substantive manner in the NRC regulatory process are maintained at—
       The NRC Technical Library
       Two White Flint North
       11545 Rockville Pike
       Rockville, MD 20852–2738

These standards are available in the library for reference use by the public. Codes and standards are usually copyrighted and may be purchased from the originating organization or, if they are American National Standards, from—
       American National Standards Institute
       11 West 42$^{nd}$ Street
       New York, NY 10036–8002
       www.ansi.org
       212–642–4900

Legally binding regulatory requirements are stated only in laws; NRC regulations; licenses, including technical specifications; or orders, not in NUREG-series publications. The views expressed in contractor-prepared publications in this series are not necessarily those of the NRC.

The NUREG series comprises (1) technical and administrative reports and books prepared by the staff (NUREG–XXXX) or agency contractors (NUREG/CR–XXXX), (2) proceedings of conferences (NUREG/CP–XXXX), (3) reports resulting from international agreements (NUREG/IA–XXXX), (4) brochures (NUREG/BR–XXXX), and (5) compilations of legal decisions and orders of the Commission and Atomic and Safety Licensing Boards and of Directors' decisions under Section 2.206 of NRC's regulations (NUREG–0750).

United States Nuclear Regulatory Commission

*Protecting People and the Environment*

NUREG/CR-7144
SAND 2008 - 3938

# Laminar Hydraulic Analysis of a Commercial Pressurized Water Reactor Fuel Assembly

Manuscript Completed:  August 2008
Date Published:  January 2013

Prepared by
E. R. Lindgren and S. G. Durbin

Sandia National Laboratory
Albuquerque, NM  87185

G. A. Zigh, NRC Project Manager

NRC Job Code Y6758

Prepared for
Division of Systems Analysis
Office of Nuclear Regulatory Research
U.S. Nuclear Regulatory Commission
Washington, DC  20555-0001

# Abstract

To the knowledge of the authors, these studies are the first hydraulic characterizations of a full length, highly prototypic 17×17 pressurized water reactor (PWR) fuel assembly in low Reynolds number flows. The advantages of full scale testing of prototypic components are twofold. First, the use of actual hardware and dimensionally accurate geometries eliminates any issues arising from scaling arguments. Second, many of the prototypic components contain intricacies by design that would not be reproduced by using simplified flow elements. While this approach yields results that are inherently specific to the fuel assembly under testing, the differences in commercial designs are considered minor, particularly when considering the hydraulics of the entire assembly.

This report summarizes the findings of the pressure drop experiments conducted using a highly prototypic PWR fuel assembly. The stated purpose of these investigations was to determine hydraulic coefficients, namely frictional loss coefficient ($S_{LAM}$) and inertial loss coefficient $\Sigma k$ values, for use in determining the hydraulic resistance in these assemblies within various numerical codes. Additionally, velocity profiles were acquired to estimate the partitioning of flow through the bundle and annular regions within the assembly. The apparatus was tested in the laminar regime with Reynolds numbers spanning from 10 to 1000, based on the average assembly velocity and hydraulic diameter.

# TABLE OF CONTENTS

# FIGURES

# TABLES

# EXECUTIVE SUMMARY

The information in this report contains hydraulic characteristics of both Pressurized Water Reactor (PWR) and Boiling Water Reactor (BWR) assemblies, which were eventually used as input data to MELCOR to perform thermal-hydraulics analysis. Among these analyses are investigations of Zirconium fire in spent fuel pool. As such, these information was considered sensitive.

The objective of this project was to provide hydraulic data to characterize the flow behavior in both the BWR and PWR assemblies. These assemblies were highly prototypical to eliminate scaling issues. Testing included pressure drop and velocity profile measurements throughout the two assemblies. These data can be further used to obtain inputs to several thermal-hydraulics codes such as MELCOR, Computational Fluid Dynamics (CFD) codes and others. These hydraulic data are useful to perform scenarios ranging from basic thermal response without accidents to severe accident cases involving Zirconium fire.

These studies found the resistance to flow through the assembly was significantly greater than predicted by generally assumed and accepted best estimate "textbook" flow parameters. Use of best estimate flow parameters may significantly underestimate the resistance to laminar flow in the assembly, which leads to an overestimate of the cooling effects of naturally induced flows that develop in dry casks under normal storage conditions or wet spent fuel pool cells during complete loss of coolant accidents. Early on, the underestimation of these parameters was balanced by the large safety margin in the evaluation of peak cladding temperature due to the low decay heat. As the applicants continue to increase the stored fuel decay heat, the correct estimation of these parameters became important to the analysis. As such, these parameters were estimated using CFD which was validated using the data in this report.

# ACKNOWLEDGEMENTS

To identify all of the individuals who participated in the success of this program would be impossible. The authors would like to acknowledge their hard work and commitment to excellence, albeit anonymously.

This work was conducted under Nuclear Regulatory Commission contract JCN# N6456. The authors gratefully acknowledge the technical guidance and support of Jorge Solis, Abdelghani Zigh, and Donald Helton at the NRC.

The authors would also like to thank Greg Koenig, Shane Adee, Glen Cannon, John Bentz, Billy Martin, Brandon Servantes, and the support staff of 6763 for their tireless efforts and dedication to service, which made the success of this project possible.

In addition, K.C. Wagner and Charles Morrow are to be commended for their outstanding technical support of the program.

# ABBREVIATIONS/DEFINITIONS

| | |
|---|---|
| AVG | average |
| BWR | boiling water reactor |
| CFD | computational fluid dynamics |
| CYBL | cylindrical boiling |
| DAQ | data acquisition |
| EPA | error propagation analysis |
| equiv. | equivalent |
| ID | inside diameter |
| IFM | intermediate fluid mixer |
| LDA | laser Doppler anemometer |
| meas. | measured |
| MELCOR | severe accident analysis code |
| NPT | national pipe thread |
| NRC | Nuclear Regulatory Commission |
| OD | outside diameter |
| PCT | peak cladding temperature |
| PWR | pressurized water reactor |
| Ref. | reference |
| RMS | root mean square |
| SFP | spent fuel pool |
| slpm | standard liters per minute (standard defined at 0°C and 1 atm) |
| SNL | Sandia National Laboratories |
| TC | thermocouple |

# 1  INTRODUCTION

## 1.1  Background

To the knowledge of the authors, these studies are the first hydraulic characterizations of a full length, highly prototypic pressurized water reactor (PWR) fuel assembly in low Reynolds number flows. The advantages of full scale testing of prototypic components are twofold. First, the use of actual hardware and dimensionally accurate geometries eliminates any issues arising from scaling arguments. Second, many of the prototypic components contain intricacies by design that would not be reproduced by using simplified flow elements. While this approach yields results that are inherently specific to the fuel assembly under testing, the differences in commercial designs are considered minor, particularly when considering the hydraulics of the entire assembly.

In a previous study under JCN Y6758 titled "Spent Fuel Pool Heatup and Propagation Phenomena Experiments", the hydraulic characterization of a full scale, highly prototypic boiling water reactor (BWR) assembly mockup was conducted using a state-of-the-art quartz crystal differential pressure gauge.[1] These pressure gauges have an unprecedented resolution of $\pm$ 0.02 $N/m^2$ ($\pm$0.000003 psi) that allows accurate measurement of the pressure drops across assembly segments at very low Reynolds numbers ($Re$ = 60 to 900).

These studies found the resistance to flow in the BWR assembly was significantly greater than predicted by generally assumed and accepted best estimate "textbook" flow parameters. Use of best estimate flow parameters may significantly underestimate the resistance to laminar flow in the assembly, which leads to an overestimate of the cooling effects of naturally induced flows that develop in dry casks under normal storage conditions or wet spent fuel pool cells during complete loss of coolant accidents. Early on, the underestimation of these parameters was balanced by the large safety margin in the evaluation of peak cladding temperature due to the low decay heat. As the applicants continue to increase the stored fuel decay heat, these parameters became important to the analysis. As such, the correct estimation of these parameters were estimated using CFD which was validated using the data in this report.

Overestimating the cooling effects of naturally induced flows may lead to non-conservative analyses of dry cask performance and spent fuel pool accident consequences. Two geometric aspects unique to the prototypic BWR assembly somewhat mitigate this problem: 1) water rods in the center of the assembly carry a significant fraction of the total natural circulation flow and aid in cooling and 2) eight of the seventy-two rods in the 9×9 BWR assembly are partial length and end 1.32 m (52 in.) below the top of the assembly. The increased void space in this upper portion greatly reduces flow resistance in this region. These geometric aspects were not fully appreciated until close inspection of the prototypic 9×9 BWR components, whose specifications are concealed in proprietary vender drawings. This is one of the benefits obtained by the Sandia National Laboratories (SNL) spent fuel pool (SFP) experimental program's use of full-scale, prototypical vender hardware.

The flow resistance issue is expected to be more acute in a pressurized water reactor (PWR) assembly. A PWR assembly is fully populated with rods along the entire length (*i.e.* no partial length fuel rods) and there are no water rods. Thus, the mitigating geometric factors found in the BWR assembly are not present in a typical PWR assembly. Furthermore, the PWR assembly has

1

more spacers, and the spacers are hydraulically longer than in a BWR assembly. All of these factors may lead to significantly higher resistance to naturally convective flow as compared to a BWR assembly. The characterization of a highly prototypic PWR assembly, especially with respect to the design and placement of the spacers, is vitally important to provide prototypic hydraulic conditions for use in the analysis of dry cask performance.

## 1.2 Objective

The objective of this project is to hydraulically characterize a full length, highly prototypic 17×17 PWR fuel assembly in low Reynolds number laminar flows expected in dry casks under normal storage conditions or wet spent fuel pool cells during complete loss of coolant accidents. Testing includes pressure drop measurements and the quantification of velocities inside the assembly.

# 2 EXPERIMENTAL APPARATUS AND PROCEDURES

These experiments were constructed and operated in the cylindrical boiling (CYBL) vessel located in Building 6585C in Technical Area III of Sandia National Laboratories (SNL) in Albuquerque, New Mexico. The CYBL facility houses a 4.9 m diameter × 8 m tall stainless steel vessel with 51 viewports. The vessel is open at the top with a 1 m diameter access port near the vessel floor. This facility was previously used for both the separate and integral effects testing in the BWR experimental program.[1]

## 2.1  Fuel Assembly

The highly prototypic fuel assembly was modeled after the 17×17 PWR. Commercial components were purchased to create the assembly including the top and bottom nozzles, spacers, intermediate fluid mixers (IFM), 24 guide tubes, one central instrumentation tube, and all related assembly hardware. Many of these components are pictured in Figure 2.1. The central instrumentation tube and guide tubes are permanently attached to the spacers to form the structural skeleton of the assembly. The top and bottom nozzles are removable. The 24 guide tubes are completely open through the top nozzle and completely blocked in the bottom nozzle. The outer diameter of the 24 guide tubes changes from 11.2 mm to 12.2 mm at an axial location 0.61 m from the bottom nozzle. Immediately above the guide tube diameter change are four 2.31 mm holes in the tube wall. The central instrumentation tube is the same diameter as the upper guide tubes and is the same diameter along the entire length. There are no holes in the instrumentation tube wall. The instrument tube is completely open through the bottom nozzle but is mostly blocked by the top nozzle with only a single 2.6 mm hole centered on the tube. If not blocked, the holes in the guide tube wall would allow some flow through the guide tubes. The vast majority of spent PWR have the guide tubes blocked by a special thimble plug or spent control rod assembly. Unless noted otherwise, the data presented in this report are for blocked guide tubes.

**Figure 2.1    Prototypic 17×17 PWR components.**

Stainless steel tubing was substituted for the fuel rod pins for hydraulic testing. The diameter of the stainless steel rods was slightly larger than prototypic pins, 9.525 mm versus 9.500 mm. Prototypic fuel rod end plugs were press fit into the ends of the stainless tubing. The dimensions of the assembly components are listed in Table 2.1.

**Table 2.1     Dimensions of assembly components in the 17×17 PWR.**

| Description | Lower Section | Upper Section |
|---|---|---|
| Number of Pins | 264 | 264 |
| Pin Diameter (mm) | 9.525 | 9.525 |
| Pin Pitch (mm) | 12.6 | 12.6 |
| Pin Separation (mm) | 3.025 | 3.025 |
| Number of Instrument Tubes | 1 | 1 |
| Number of Guide Tubes (G/T) | 24 | 24 |
| G/T Diameter (mm) | 11.2 | 12.2 |
| Axial Length (m) | 0.704 | 3.268 |

Figure 2.2 shows the rod configuration of a typical 17×17 PWR assembly. Three different storage cells characterized by an inner dimension $D_{cell}$ were studied. These cells are discussed next.

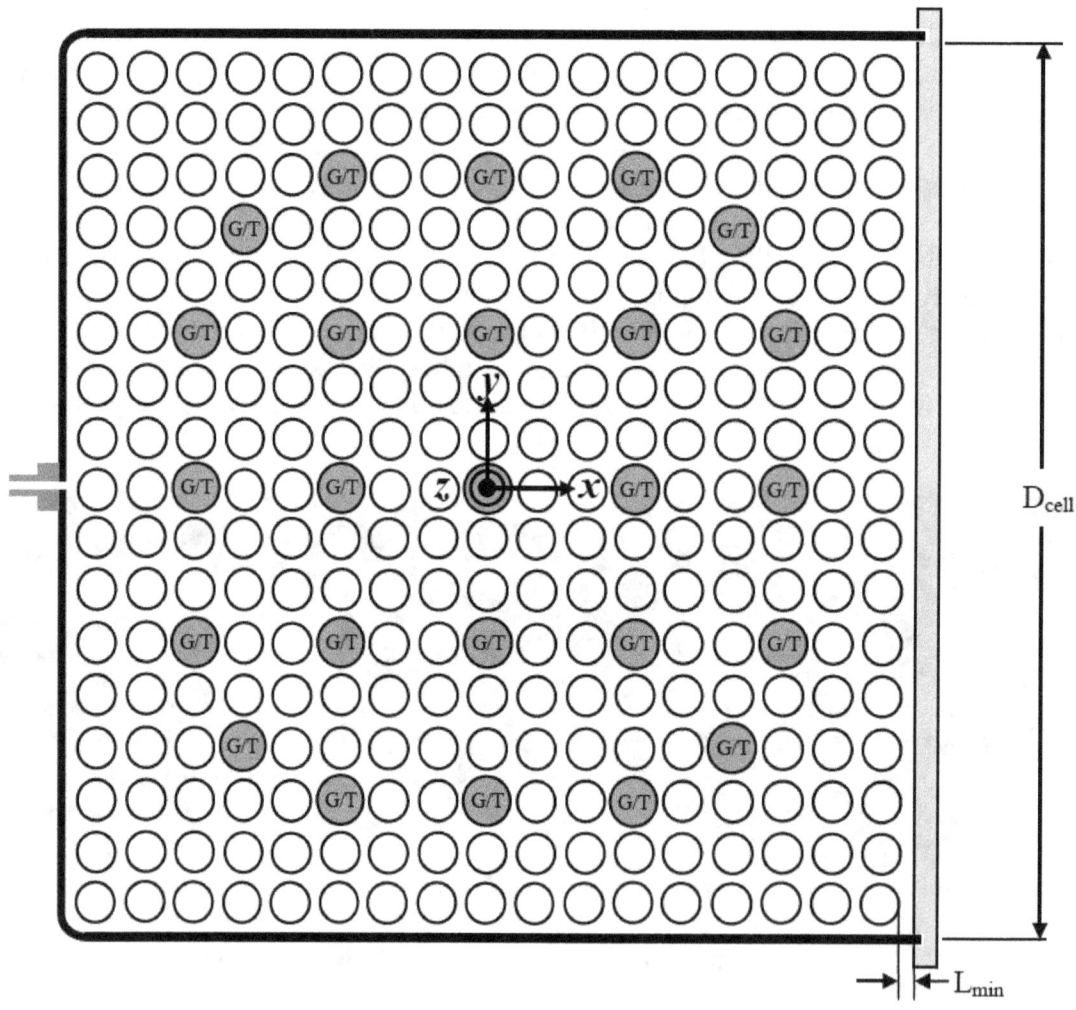

**Figure 2.2     Rod configuration of a typical 17×17 PWR assembly.**

## 2.2    Storage Cells

A major difference between BWR and PWR assemblies is the absence of the channel box in the PWR. Without a channel box component, the pool or cask storage cell defines the peripheral flow boundaries for the PWR assembly (*i.e.*, similar to a BWR canister). The extent of the gap between the outer row of rods and the inside cell wall influences the nature of flow inside the bundle. In order to study this effect, three different sized storage cells were tested. The three sizes were chosen to represent two common commercial sizes and one small cell size that minimized the annular flow much like the BWR canister. Table 2.2 lists the dimensions of cells used in three Holtec dry casks and one pool rack. The outer dimension of the 17×17 spacer is 214.0 mm (8.43 in.) and represents the smallest possible storage cell that would fit on the assembly. The internal dimensions of the commercial cells range considerably from a low of 222.2 mm (8.75 in.) to a high of 229.9 mm (9.05 in.).

Since there is variation in the size of the cells used commercially, three different sized pool cells were tested as indicated in gray in Table 2.2 and depicted in Figure 2.2 as dimension $D_{cell}$. The base case was the 217.5 (8.56 in.) cell with minimal annular flow area. Although this size is not found commercially, it is an important case for comparison purposes as it minimizes the complexity of the annular gap flow. The two other experimental storage cell sizes are 221.8 mm (8.73 in.) similar to that found in the Holtec MPC-24E/EF cask and 226.6 mm (8.92 in.) like found in the Holtec MPC-24 cask. This span of sizes includes the size used in a typical Holtec spent fuel pool rack.

Table 2.2    Internal dimensions of commercial cask, pool cells, and areas of as-built experimental cells.

| | Number in Cask | Length (mm) | Width (mm) | Experimental Storage Cell Dimension, $D_{cell}$ (mm) | Minimum Annular Gap, $L_{min}$ (mm) |
|---|---|---|---|---|---|
| 17×17 spacer | - | 214.0 | 214.0 | 217.5 | 1.75 |
| | | | | | |
| MPC-32 | 22 | 223.3 | 223.3 | | |
| | 8 | 223.3 | 227.1 | | |
| | 2 | 227.1 | 227.1 | | |
| | | | | | |
| MPC-24E/EF | 20 | 222.2 | 222.2 | 221.8 | 3.9 |
| | 4 | 229.9 | 229.9 | | |
| | | | | | |
| MPC-24 | 24 | 226.6 | 226.6 | 226.6 | 6.3 |
| | | | | | |
| Holtec pool | - | 224.8 | 224.8 | | |

## 2.3    Air Flow Control

Air was metered into the bottom of the assembly with eight mass flow controllers (MKS Instruments Inc. Model 1559A). The upper flow ranges in standard liters per minute (slpm) of each controller are listed in the table inset in Figure 2.3. Figure 2.3 also shows a diagram of the flow metering and flow straightening components as well as photographs of the equipment used. The metered air was conditioned to produce a uniform velocity profile at the inlet to the assembly. The flow from all the flow controllers was routed through a baffled manifold to mix

5

the flows and equally distribute flow to the bottom of a square flow straightener. The flow straightener consisted of a course screen, followed by a plastic honeycomb, followed by a fine screen. The selection of straightening components and their placement are based on the results of Farell and Youssef.[2]

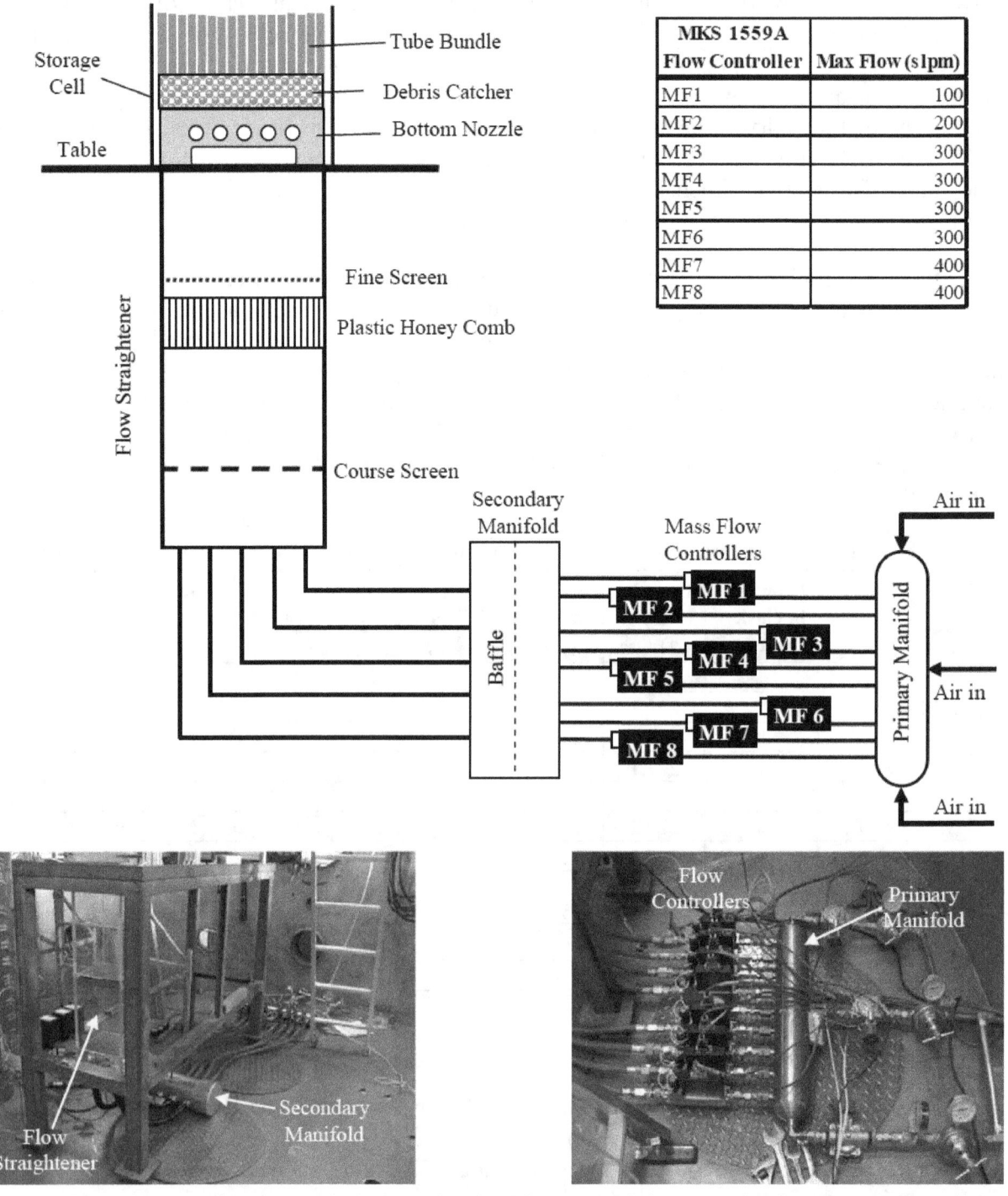

| MKS 1559A Flow Controller | Max Flow (slpm) |
|---|---|
| MF1 | 100 |
| MF2 | 200 |
| MF3 | 300 |
| MF4 | 300 |
| MF5 | 300 |
| MF6 | 300 |
| MF7 | 400 |
| MF8 | 400 |

Figure 2.3    Diagram and photographs of the flow control and flow straightening systems.

## 2.4    Pressure Drop Measurements

Figure 2.4 shows the layout of the PWR pressure drop experimental assembly, including all available pressure port locations. Three Paroscientific Digiquartz differential pressure transducers (Model 1000-3D) were plumbed directly to the desired pressure ports.  These pressure gauges use a highly sensitive quartz crystal to measure slight changes in differential pressure (resolution ~0.02 Pa).  For this report the pressure drop across two ports is reported as the low side port first followed by the high side port, *e.g.*, the overall pressure drop across the entire assembly is reported as A–36 (where "A" denotes ambient).  Unless noted otherwise, the measurements described hereafter were taken with the guide tubes blocked.  These guide tubes were blocked at the top, which is open, by inserting tapered rubber plugs into each tube.  Figure 2.5 shows the lower section of the guide tube including the transition in the outer diameter and the location of the drain holes.

Measurements were recorded directly to the hard drive of a PC-based data acquisition system every 2 seconds using a LabView 8.2 interface.  These measurements included the air flow rate through the assembly, ambient air temperature, ambient air pressure, and the assembly pressure drops.  The LabView interface was used to automatically change the air flow rate according to a prescribed program which allowed a greater number of flow rate settings than in previous testing.

With each pressure transducer plumbed to two set port locations and with the air flow off, pressure drop measurements were recorded for a period of roughly 1 minute. These measurements were termed zero flow measurements and allowed for correction of any zero drift in the transducer.  Next, the air flow was set to the desired rate with pressure drop readings subsequently acquired for 2 minutes. The air flow was then stopped, and zero flow measurements were again taken for 1 minute. This procedure was repeated for different air flow rates.  A typical set of pressure traces are shown in                          Figure 2.6.  The pressure spikes in this figure evident during the reestablishment of flow are discarded before averaging for the pressure drops.  Also, the slight zero drift of the transducer was corrected by subtracting the average of the zero flow measurements taken prior to and after each respective flow test.  The zero corrections of the pressure drops were less than 0.92 $N/m^2$, which occurred during an overall A–36 pressure drop measurement.

Additionally, the BWR characterized in a previous study documented in SAND2007-2270 was re-tested using the improved automated procedures described previously.[1]  The pressure port locations for the BWR assembly are shown in Figure 2.7.

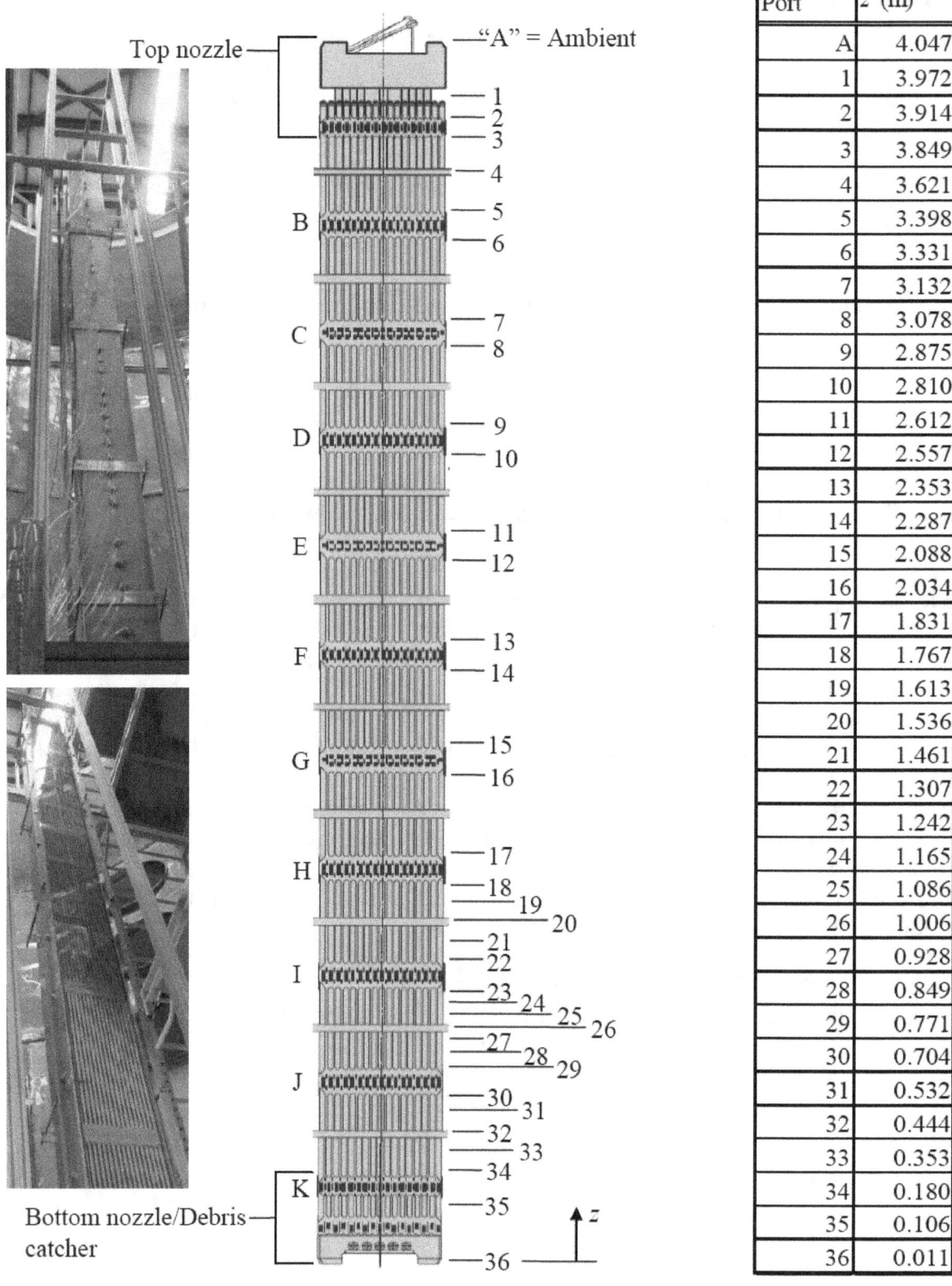

| Port | $z$ (m) |
|------|---------|
| A | 4.047 |
| 1 | 3.972 |
| 2 | 3.914 |
| 3 | 3.849 |
| 4 | 3.621 |
| 5 | 3.398 |
| 6 | 3.331 |
| 7 | 3.132 |
| 8 | 3.078 |
| 9 | 2.875 |
| 10 | 2.810 |
| 11 | 2.612 |
| 12 | 2.557 |
| 13 | 2.353 |
| 14 | 2.287 |
| 15 | 2.088 |
| 16 | 2.034 |
| 17 | 1.831 |
| 18 | 1.767 |
| 19 | 1.613 |
| 20 | 1.536 |
| 21 | 1.461 |
| 22 | 1.307 |
| 23 | 1.242 |
| 24 | 1.165 |
| 25 | 1.086 |
| 26 | 1.006 |
| 27 | 0.928 |
| 28 | 0.849 |
| 29 | 0.771 |
| 30 | 0.704 |
| 31 | 0.532 |
| 32 | 0.444 |
| 33 | 0.353 |
| 34 | 0.180 |
| 35 | 0.106 |
| 36 | 0.011 |

**Figure 2.4**      **Experimental PWR apparatus showing as-built port locations.**

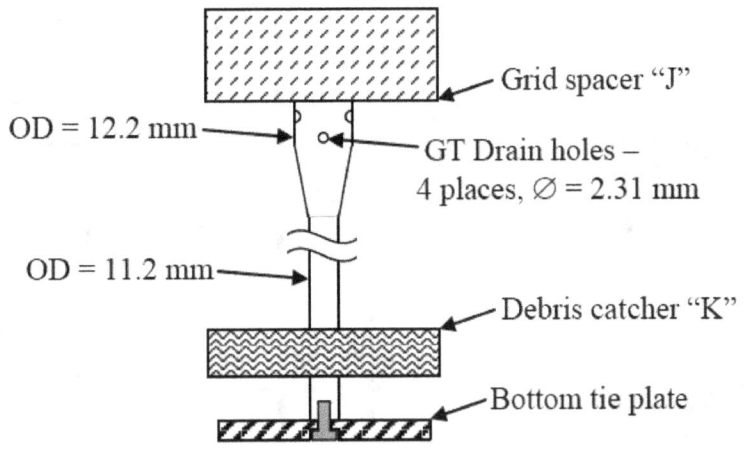

Figure 2.5        Lower detail of the PWR guide tube.

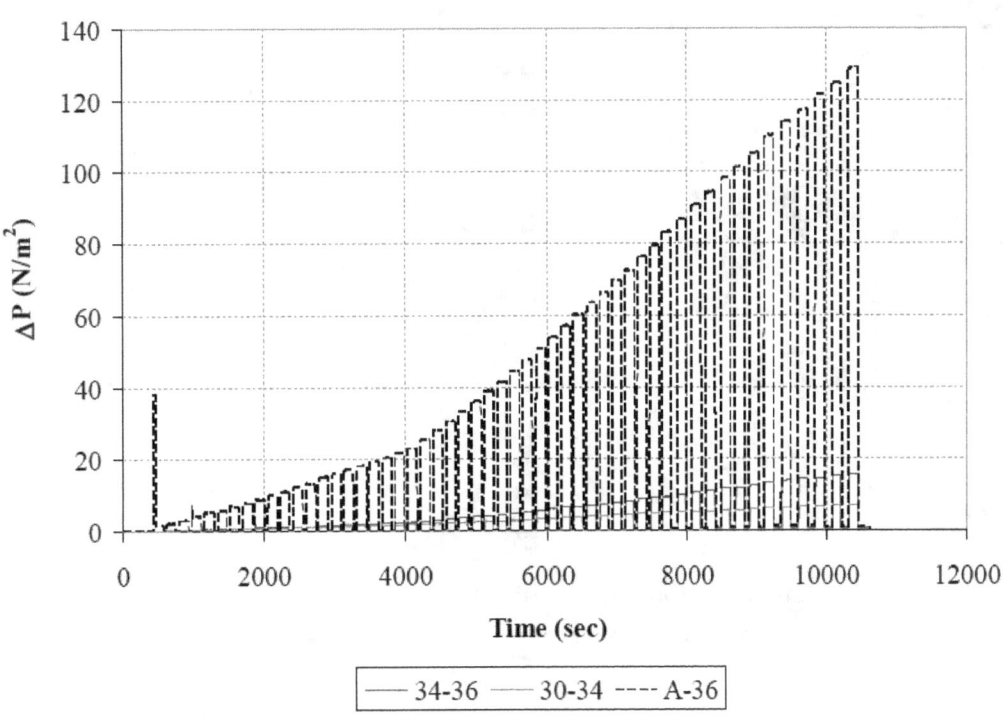

Figure 2.6        Pressure traces recorded during a typical flow profile for measurements
across 34–36, 30–34, and A–36.

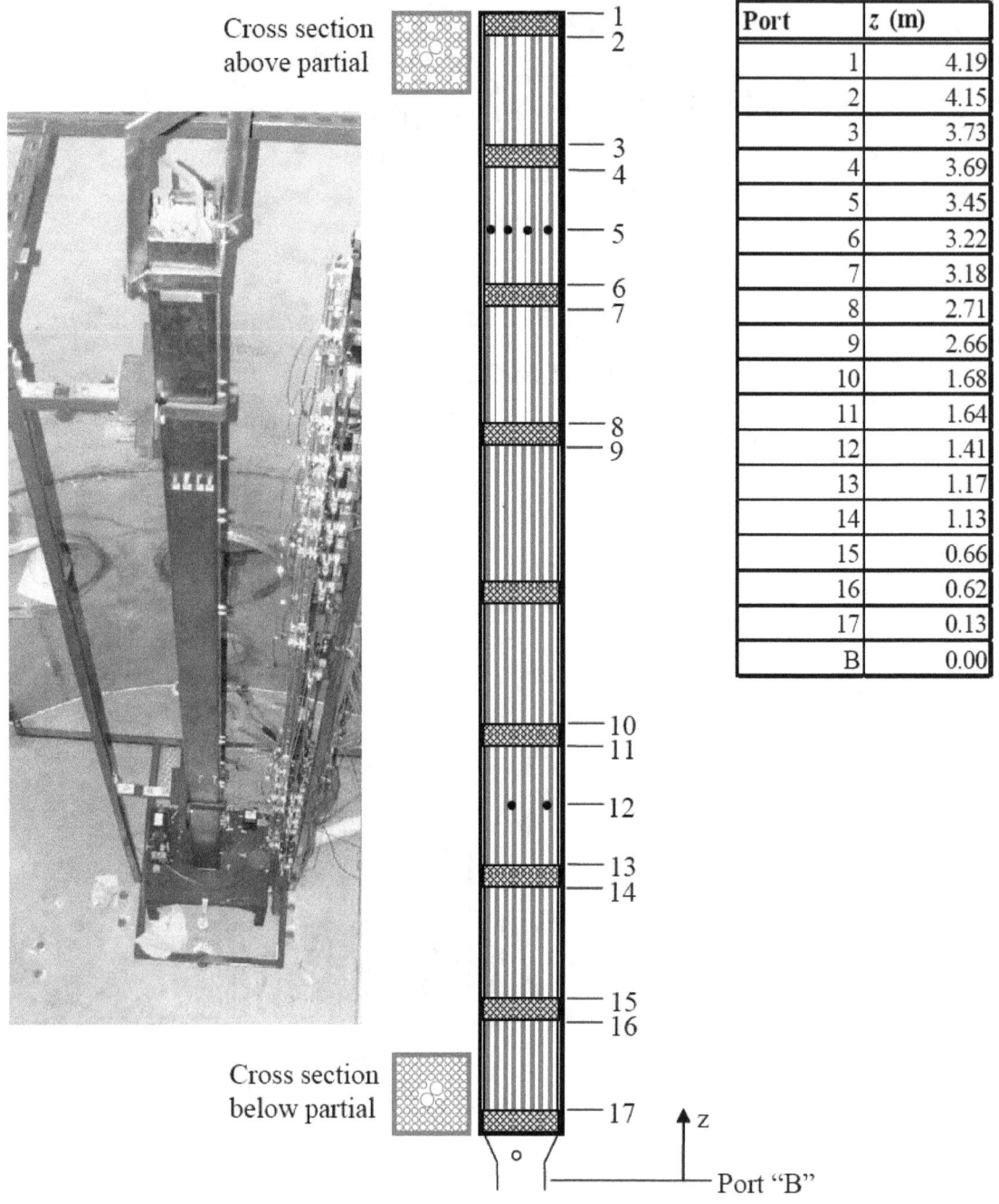

| Port | z (m) |
|------|-------|
| 1 | 4.19 |
| 2 | 4.15 |
| 3 | 3.73 |
| 4 | 3.69 |
| 5 | 3.45 |
| 6 | 3.22 |
| 7 | 3.18 |
| 8 | 2.71 |
| 9 | 2.66 |
| 10 | 1.68 |
| 11 | 1.64 |
| 12 | 1.41 |
| 13 | 1.17 |
| 14 | 1.13 |
| 15 | 0.66 |
| 16 | 0.62 |
| 17 | 0.13 |
| B | 0.00 |

**Figure 2.7** **Experimental BWR apparatus showing as-built port locations.**

### 2.4.1 Hydraulic Loss Coefficients

As discussed previously, one goal of this research was to determine the $S_{LAM}$ and $\Sigma k$ coefficients for use with thermal-hydraulic codes. The cross-sectional areas and hydraulic diameters used in the hydraulic analyses to follow are presented in    Table 2.3. These reference hydraulics are computed from the bundle cross sections in the assemblies and do not account for contractions encountered at spacers and nozzle/tie plate elements. The reference hydraulic diameter is defined according to convention, or four times the flow area divided by the wetted perimeter.

10

**Table 2.3** Summary of the reference flow areas and hydraulic diameters of the laminar flow characterization experiments.

| Description | Storage Cell Dimension, $D_{cell}$ (mm) | Flow Area, $A_{Assembly}$ ($m^2$) | $D_{H, Ref}$ (m) |
|---|---|---|---|
| PWR Upper Bundle | 217.5 | 0.0256 | 0.0105 |
| PWR Lower Bundle | 217.5 | 0.0260 | 0.0108 |
| PWR Upper Bundle | 221.8 | 0.0275 | 0.0113 |
| PWR Lower Bundle | 221.8 | 0.0279 | 0.0116 |
| PWR Upper Bundle (Unblocked Guide Tubes) | 221.8 | 0.0299 | 0.0113 |
| PWR Lower Bundle (Unblocked Guide Tubes) | 221.8 | 0.0279 | 0.0116 |
| PWR Upper Bundle | 226.6 | 0.0296 | 0.0121 |
| PWR Lower Bundle | 226.6 | 0.0301 | 0.0124 |
| BWR Upper (Partially Populated) | 132.6 | 0.0106 | 0.0141 |
| BWR Lower (Fully Populated) | 132.6 | 0.0098 | 0.0119 |

Curve fits to the pressure drop data were used to determine the $S_{LAM}$ and $\Sigma k$ coefficients of the assembly. The technique used to determine these coefficients was successfully validated by investigation of flow in a simple annulus for which an analytic value of $S_{LAM}$ is known, see Appendix A for details. The determination of these coefficients is discussed next. The major, or viscous, pressure loss is expressed in Equation 1 as a function of the average $z$-component of velocity in the assembly, $W_{Assembly}$.

$$\Delta P_{major} = f \left( \frac{L}{D_{H, Ref}} \right) \left( \frac{\rho \cdot W_{Assembly}^2}{2} \right) \qquad 1$$

The friction factor for laminar flow is written explicitly as

$$f = \frac{S_{LAM}}{Re}, \text{ where } S_{LAM} = 64 \text{ (pipe flow)}$$
$$= 100 \text{ (bundle flow)} \qquad 2$$

Reynolds number is defined for these studies as

$$Re = \frac{\rho \cdot W_{Assembly} \cdot D_{H, Ref}}{\mu} \qquad 3$$

Substituting for the Reynolds number yields

$$\Delta P_{major} = S_{LAM} \left( \frac{L}{D_{H, Ref}^2} \right) \left( \frac{W_{Assembly} \cdot \mu}{2} \right) \qquad 4$$

The minor, or form, pressure drops across the assembly are given by

$$\Delta P_{minor} = \sum k \left( \frac{\rho \cdot W_{Assembly}^2}{2} \right) \qquad \textbf{5}$$

Curve fits to pressure drop data are presented in the following format. In Equation 6, the quadratic term accounts for the minor losses and the linear term for the major losses.

$$\Delta P_{total} = a_2 \cdot W_{Assembly}^2 + a_1 \cdot W_{Assembly} \qquad \textbf{6}$$

The assembly velocity is determined from the measurement of the volumetric flow of air through the assembly as detailed in Section 2.3. The assembly velocity is computed by the following conversion in Equation 7. The standard density is taken at 0°C and 1 atm. The measured density is determined from local temperature and atmospheric pressure using the ideal gas law.

$$W_{Assembly} \, (m/s) = \frac{Q_{tot} \, (slpm) \cdot \left( \rho_{std} / \rho_{meas} \right)}{A_{Assembly} \, (m^2)} \cdot \left( \frac{1 \, m^3}{1000 \, liters} \right) \cdot \left( \frac{1 \, min}{60 \, sec} \right) \qquad \textbf{7}$$

Because the total pressure drop is simply the sum of the major and minor pressure drops, the $S_{LAM}$ and $\Sigma k$ coefficients may now be determined explicitly.

$$S_{LAM} = 2 \cdot a_1 \left( \frac{D_{H, Ref}^2}{L \mu} \right)$$

$$\sum k = \frac{2 \cdot a_2}{\rho} \qquad \textbf{8}$$

## 2.5 Laser Doppler Anemometry Measurements

Laser Doppler anemometry (LDA) is a non-intrusive, optical technique used to measure the instantaneous velocity in a flow field at the intersection of two coherent laser beams. The most common method of LDA used at present is the dual-beam anemometer system. Typically, a single laser beam is split into two mutually coherent polarized light waves, which intersect to form a spheroid-shaped region called the measuring volume. Particles passing through the measuring volume with a given velocity scatter light from the light beams to produce the LDA signal. For the investigations detailed herein, a dual-beam system operating in backscatter mode as shown in Figure 2.8 was used. The photodetector was a photomultiplier (PM), and a Bragg cell was used to introduce a reference frequency, allowing the measurement of near-zero velocities. The signal was then processed in a burst analyzer and sent to a PC-based data acquisition system.

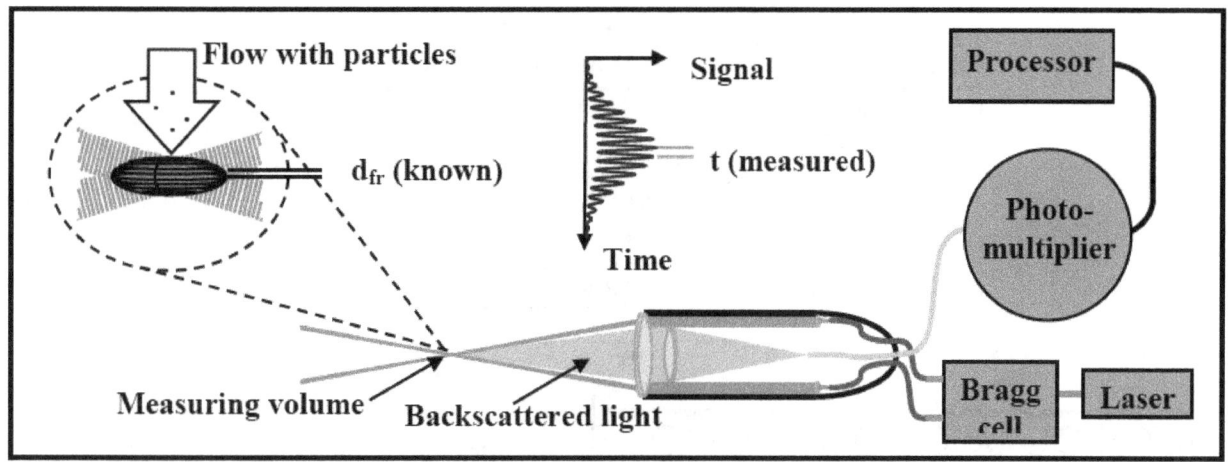

**Figure 2.8**     **Dual-beam backscatter LDA system components and principles.**

A visual interpretation of the operation of the dual-beam LDA is given by the fringe model. This model avoids reference to the Doppler shift effect and instead makes use of the interference fringes created in the probe volume by the crossing of two incident coherent light beams. The wave fronts of the two beams form interference fringes with spacing $d_{fr}$. The bright fringes are created from the constructive interference of the incident light beams and the dark fringes are a result of destructive interference. The fringe spacing $d_{fr}$ is a function of the half angle of the beam intersection and the wavelength of the incident light beams. The velocity of a particle passing perpendicular between fringes is then the division of the fringe spacing and the time between subsequent signal peaks. This model represents an oversimplification of the actual physics and signal processing involved in the LDA technique but is presented to give the unaccustomed reader some understanding of the instrument capability and functionality. Further information on the LDA technique can be found in Durst and Melling as well as Albrecht, *et al.*[3,4]

### 2.5.1   LDA Experimental Setup

The average velocity and root-mean-square (RMS) velocity fluctuations were measured in these experiments by a single-component laser-Doppler anemometer (Dantec 1-D FlowExplorer). This LDA system is composed of the FlowExplorer probe head, a photomultiplier, a burst analyzer, a motorized stage and controller unit, a PC-based data acquisition (DAQ) system, and data processing software. These components are listed in Table 2.4.

Figure 2.9 shows the layout of the test components for LDA measurements. The LDA probe is mounted externally to the PWR assembly on a motorized stage. The laser beams pass through the optical window into the assembly and measure the velocity at the intersection of the beams. In this manner the local velocity can be measured across the assembly in between rod banks. Figure 2.10 gives two photographs of the LDA setup. These photographs depict a measurement just inside the optical window.

Table 2.4    Detailed list of the LDA system components.

| Description | Manufacturer | Serial Number |
|---|---|---|
| FlowExplorer probe head | Dantec Dynamics | 0115 |
| BSA F60 – Burst analyzer | Dantec Dynamics | 437 |
| Photomultiplier | Dantec Dynamics | 119 |
| 1-D Stage | Isel Automation | 505 |
| 1-D Stage controller | Isel Automation | 502 |
| DAQ – PC | Dell | F51KYD1 |

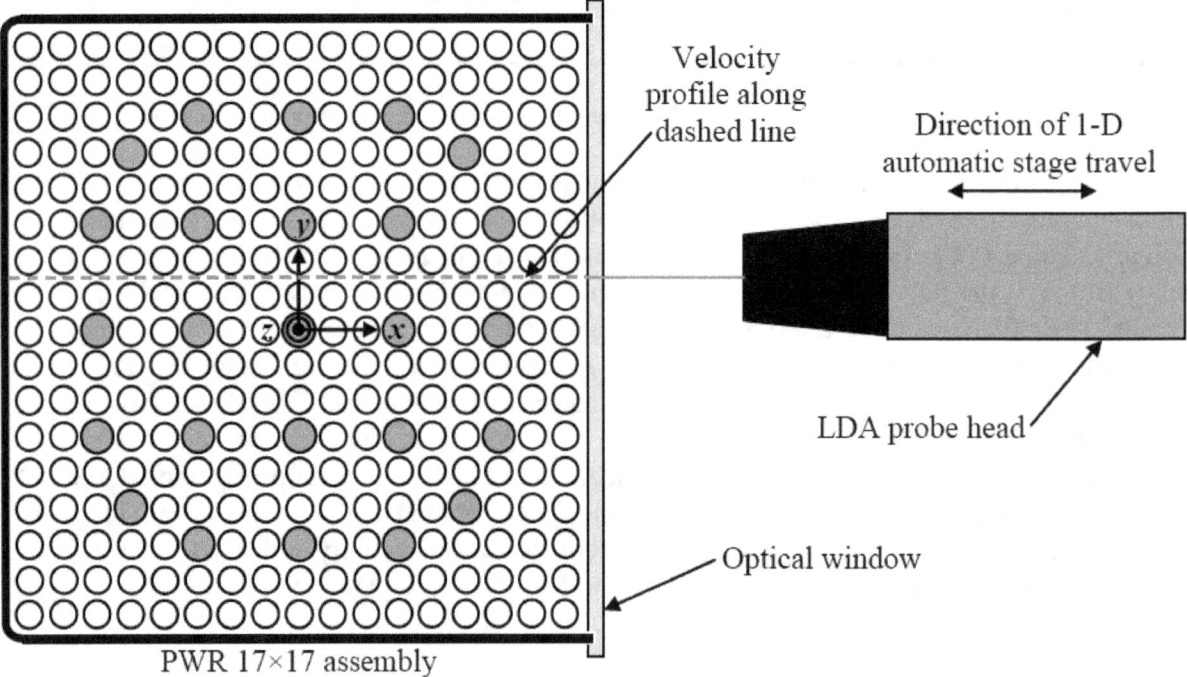

Figure 2.9    Schematic of the LDA system for measuring velocity profiles in the PWR 17×17 assembly.

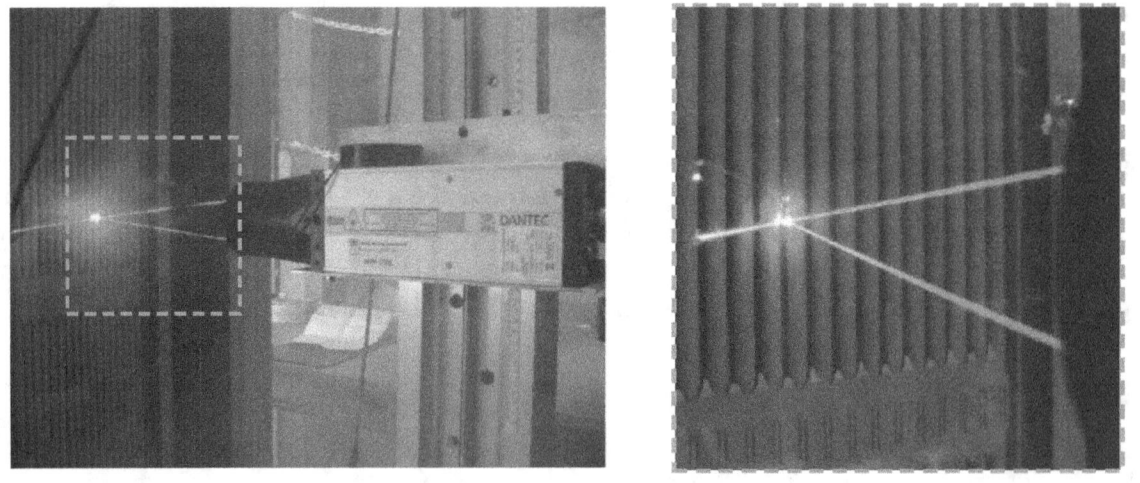

Figure 2.10    Photographs showing the LDA probe system in relation to the fuel assembly.

# 3    PRESSURE DROP RESULTS

## 3.1    217.5 mm Storage Cell

Since there is variation in the size of the cells used commercially for storing spent PWR assemblies, three different sized storage cells were tested. The base case is the 217.5 (8.56 in.) cell with minimal annular flow area. Although this size is not found commercially, this size cell is an important case for comparison purposes as it minimizes the complexity of the annular gap flow. The base case PWR storage cell is analogous to the BWR canister.

### 3.1.1    Overall Pressure Drop

#### 3.1.1.1    Overall Pressure Drop Dependence on Reynolds Number

Figure 3.1 shows the overall assembly pressure drop as a function of Reynolds number in the bundle of the PWR (A–36) and BWR (1–B) assemblies. The Reynolds number for the PWR is determined from the upper bundle, reference hydraulic diameter and the assembly velocity. The Reynolds number for the BWR is determined from the lower bundle, reference hydraulic diameter and the assembly velocity. The BWR data is added for comparison and includes the new data collected in the present study along with the data collected in the previous study.[1] The PWR assembly is contained inside of the smallest, 217.5 mm storage cell that is sized to be analogous to the BWR canister. The water rods in the BWR and the guide tubes in the PWR are blocked to flow. At all Reynolds numbers, the pressure drop across the PWR is significantly larger than across the BWR assembly.

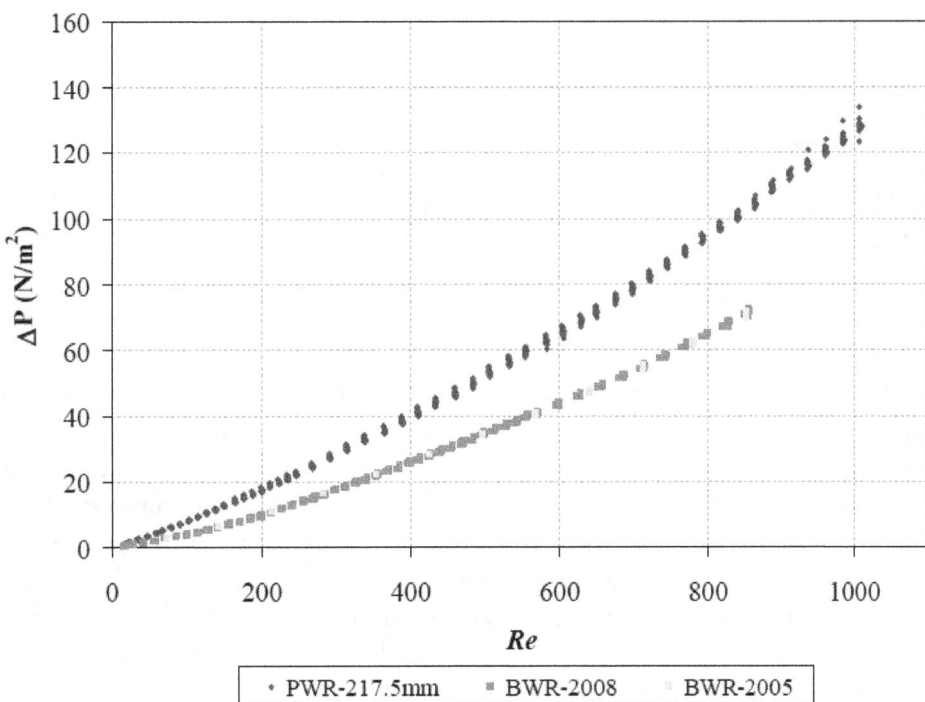

Figure 3.1    Assembly pressure drop as a function of Reynolds numbers for analogous PWR and BWR fuel bundles.

### 3.1.1.2  Pressure Drop with Axial Position

Figure 3.2 shows the axial pressure drop in a PWR assembly for three different flows when placed inside the smallest storage cell. The size of this storage cell is comparable to the canister on a BWR assembly. The pressure drop measured in the BWR assembly is shown for comparison. The pressure drop in the PWR is significantly higher than in the BWR assembly. The pressure drop along the PWR bundle and across the PWR spacers is slightly greater than the corresponding locations in the fully populated lower section of the BWR assembly and significantly greater than in the upper partially populated section of the BWR assembly.

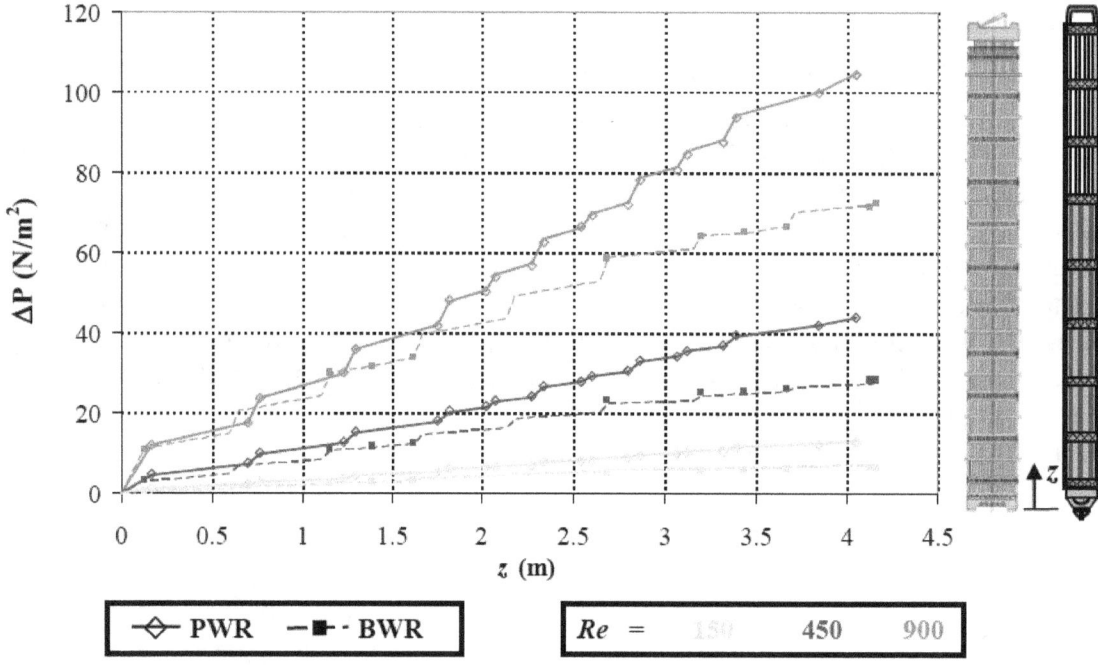

**Figure 3.2**   Assembly pressure drop as a function of axial height for analogous PWR and BWR fuel bundles.

### 3.1.2   Hydraulic Loss Coefficients

### 3.1.2.1   Full Flow Range $S_{LAM}$-$\Sigma k$ Analysis

The following is an example $S_{LAM}$-$\Sigma k$ analysis of the curve fit to pressure drop data for the full range of laminar flows. Please refer to Figure 2.4 for the location of the pressure ports described next. The data in Figure 3.3 refer to the pressure drops across pressure ports 34–36, 30–34, and A–36 for the assembly inside the 217.5 mm storage cell. The following analysis assumes that the hydraulic loss coefficients are not significant functions of the laminar flow Reynolds number. In the Section 3.1.2.2, the same data are analyzed in a partitioned fashion to show the Reynolds number dependence. The technique used to determine the hydraulic loss coefficients was successfully validated by investigation of flow in a simple annulus for which an analytic value for $S_{LAM}$ is known, see Appendix A for details.

Note: The pressure drops are plotted versus the corresponding assembly velocity for each section. The reference hydraulic diameter and flow area of the upper portion of the assembly were used to calculate $S_{LAM}$ and $\Sigma k$ coefficients for the A–36 data, since these hydraulically characterize the bulk of the assembly.

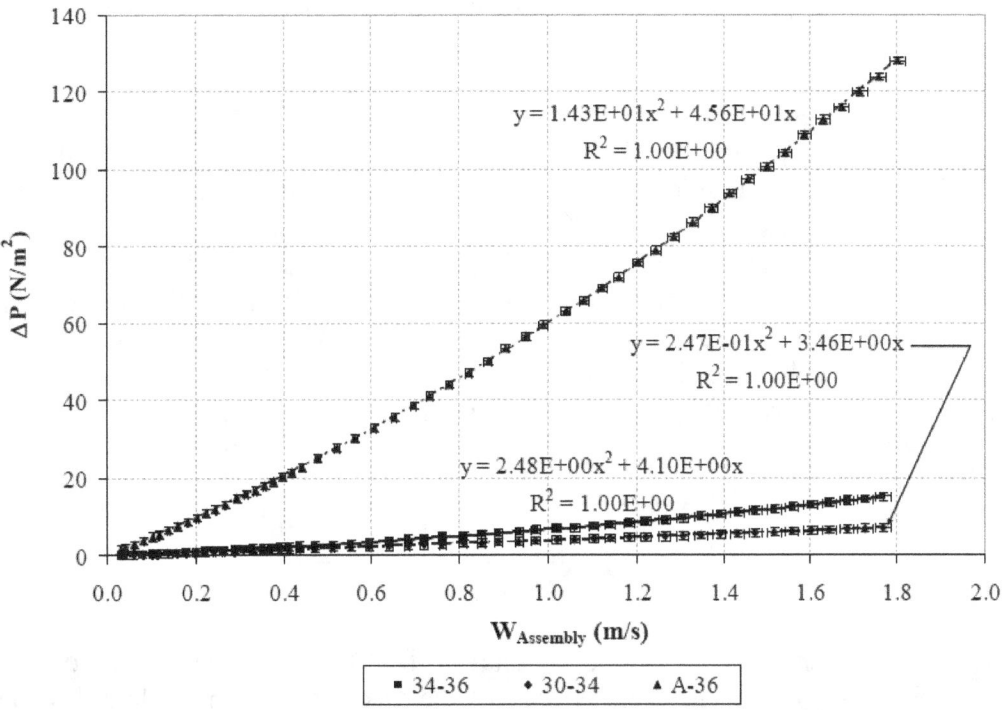

Figure 3.3    Pressure drop as a function of average assembly velocity for the PWR fuel mockup inside the 217.5 mm storage cell.

The relevant information for the calculation of the overall $S_{LAM}$ and $\Sigma k$ coefficients are given in Table 3.1. Uncertainties in the $S_{LAM}$ and $\Sigma k$ coefficients listed in this report are taken as ±5 and ±1.4, respectively. See Appendix B for details.

Table 3.1    $S_{LAM}$ and $\Sigma k$ coefficient analysis data for pressure drops between 34–36, 30–34, and A–36 for the 217.5 mm storage cell.

| Pressure Drop | L (m) | $A_{Assembly}$ (m$^2$) | $D_{H, Ref.}$ (m) | $a_1$ (N·s/m$^3$) | $a_2$ (N·s$^2$/m$^4$) | $S_{LAM}$ | $\Sigma k$ |
|---|---|---|---|---|---|---|---|
| 34–36 | 0.1688 | 0.0260 | 0.0108 | 4.10 | 2.48 | 305 | 5.0 |
| 30–34 | 0.5244 | 0.0260 | 0.0108 | 3.46 | 0.247 | 83.0 | 0.50 |
| A–36 | 4.0472 | 0.0256 | 0.0105 | 45.6 | 14.3 | 135 | 29 |

This analysis assumes air properties at local ambient conditions, typically $\rho = 0.98$ kg/m$^3$ and $\mu = 1.85 \times 10^{-5}$ N·s/m$^2$. Changes in air temperature and pressure are taken into account for measurements collected during different experimental runs.

A summary of the $S_{LAM}$ and $\Sigma k$ coefficients for the 217.5 mm storage cell testing is shown in Table 3.2. These values were determined from the full experimental flow rate range of 30 to 2100 slpm, or Reynolds numbers of 10 to 1000, respectively. The reference hydraulic diameter and flow area of the upper section was used to calculate the $S_{LAM}$ and $\Sigma k$ values for any span including the upper section, *e.g.*, A–36. Values of $S_{LAM}$ and $\Sigma k$ for segments 34–36 and 30–34 are calculated in two ways: 1) using the reference hydraulics of upper bundle section ($D_{H, Ref.}$ = 0.0105 m) and 2) using the reference hydraulics of lower bundle section ($D_{H, Ref.}$ = 0.0108 m). The second set of $S_{LAM}$ and $\Sigma k$ represent the actual local hydraulic conditions. The first set of $S_{LAM}$ and $\Sigma k$ parameters are needed to appropriately integrate the segment results for comparison with the overall (A–36) results as described next.

The single section/spacer $S_{LAM}$ and $\Sigma k$ values can be manipulated to recreate the values observed in the multiple section/spacer data. To accomplish this, the values of $S_{LAM}$ and $\Sigma k$ must be calculated with the same hydraulic parameters (*i.e.*, $D_{H, Ref.}$ and flow area) as used in the overall analysis. Additionally, $S_{LAM}$ values must be weight averaged based on flow length. Equation 9 shows the general format for calculating the effective $S_{LAM}$ coefficient of an assembly span with an overall flow length of $L_{tot}$, "I" number of spacers (including flow nozzles and IFMs), and "J" number of bundle sections. This effective $S_{LAM}$ coefficient is tabulated near the lower right hand corner of Table 3.2. The directly observed value of $S_{LAM}$ for the overall pressure drop was 132.9, which denotes an error of less than 0.2% from the length-averaged value of 132.7.

$$S_{LAM,eff} = \frac{\left( \sum_{i=1}^{I} L_{sp,i} \cdot S_{LAMsp,i} + \sum_{j=1}^{J} L_{sect,j} \cdot S_{LAMsect,j} \right)}{L_{tot}} \qquad 9$$

The form loss coefficients will simply be the total of the $\Sigma k$ values for the individual segments. This value of 30.7 is provided near the bottom of the "$\Sigma k$" column in Table 3.2. The directly measured value of $\Sigma k$ for the overall assembly pressure drop was 30.6, which indicates an error of about 0.3%. These two comparisons provide a measure of assurance that the values of $S_{LAM}$ and $\Sigma k$ presented here provide a consistent set of data.

**Table 3.2    Full flow range $S_{LAM}$ and $\Sigma k$ coefficients for the PWR assembly in the 217.5 mm storage cell.**

| Segment | Description | $D_{H, Ref.}$ (m) | $S_{LAM}$ | $\Sigma k$ | L (m) | $S_{LAM} \cdot (L/L_{tot})$ |
|---|---|---|---|---|---|---|
| A–3 | Top Nozzle | 0.0105 | 107.1 | 1.5 | 0.2097 | 5.5 |
| 3–5 | Long Bundle | 0.0105 | 76.1 | 1.4 | 0.4509 | 8.5 |
| 5–6 | Spacer | 0.0105 | 418.4 | 2.0 | 0.0667 | 6.9 |
| 6–7 | Short Bundle | 0.0105 | 82.4 | 1.0 | 0.1990 | 4.1 |
| 7–8 | IFM | 0.0105 | 283.2 | 1.4 | 0.0545 | 3.8 |
| 8–9 | Short Bundle | 0.0105 | 77.0 | 0.4 | 0.2021 | 3.8 |
| 9–10 | Spacer | 0.0105 | 466.3 | 1.9 | 0.0656 | 7.6 |
| 10–11 | Short Bundle | 0.0105 | 81.1 | 0.8 | 0.1979 | 4.0 |
| 11–12[*] | IFM | 0.0105 | 290.8 | 1.4 | 0.0545 | 3.9 |
| 12–13[*] | Short Bundle | 0.0105 | 77.2 | 0.7 | 0.2043 | 3.9 |
| 13–14 | Spacer | 0.0105 | 432.1 | 2.0 | 0.0661 | 7.1 |
| 14–15 | Short Bundle | 0.0105 | 72.1 | 0.8 | 0.1995 | 3.6 |
| 15–16 | IFM | 0.0105 | 294.2 | 1.5 | 0.0534 | 3.9 |
| 16–17 | Short Bundle | 0.0105 | 73.3 | 0.3 | 0.2027 | 3.7 |
| 17–18 | Spacer | 0.0105 | 492.5 | 1.9 | 0.0646 | 7.9 |
| 18–22 | Long Bundle | 0.0105 | 80.7 | 1.0 | 0.4604 | 9.2 |
| 22–23 | Spacer | 0.0105 | 480.0 | 1.6 | 0.0646 | 7.7 |
| 23–29 | Long Bundle | 0.0105 | 77.2 | 1.5 | 0.4710 | 9.0 |
| 29–30 | Spacer | 0.0105 | 424.3 | 2.2 | 0.0667 | 7.0 |
| 30–34 | Long Bundle | 0.0105 | 77.4 | 0.5 | 0.5244 | 10.0 |
| 30–34 | Long Bundle | 0.0108 | 83.0 | 0.5[†] | -- | -- |
| 34–36 | Bottom Nozzle | 0.0105 | 284.7 | 4.9 | 0.1688 | 11.9 |
| 34–36 | Bottom Nozzle | 0.0108 | 305.3 | 5.0[†] | -- | -- |
| Summation | Overall (equiv.) | 0.0105 | -- | 30.7 | 4.0472 | 132.7 |
| A–36 | Overall (meas.) | 0.0105 | 132.9 | 30.6 | 4.0472 | -- |

[*] Due to a blockage in port 12, segments 11–12 and 12–13 values are averages of the remaining IFM spacers and short bundle segments, respectively.
[†] These form losses are not included in the sum total.

### 3.1.2.2  Partitioned Flow Range $S_{LAM}$-$\Sigma k$ Analysis

The previous analysis assumed that the hydraulic loss coefficients are not significant functions of the laminar flow Reynolds number. In this section, the same overall assembly (A–36) pressure drop data are analyzed in a partitioned fashion to show the Reynolds number dependence. The data were divided into three partitions of approximately equal number of data points. Each partition of data was analyzed for $S_{LAM}$ and $\Sigma k$ using the same approach described in the previous section.

Figure 3.4 shows the $S_{LAM}$ and $\Sigma k$ for the three-partition analysis of data for the 217.5 mm cell. The points represent the $S_{LAM}$ and $\Sigma k$ for the corresponding average Reynolds number for the data included in the partition. The horizontal bars represent the range of Reynolds numbers over which the hydraulic loss coefficients were calculated. The vertical bars represent the error for the $S_{LAM}$ and $\Sigma k$ calculated by the method detailed in Appendix B. The error is large for small Reynolds numbers and decreases as the Reynolds number increases. The dashed black lines represent the overall $S_{LAM}$ and $\Sigma k$ values reported in Table 3.2. The yellow shading about the black dashed lines represents the experimental error associated with the overall $S_{LAM}$ and $\Sigma k$

values. The error range for all the partitioned data overlaps with the error range of the overall $S_{LAM}$ and $\Sigma k$. The hydraulic loss coefficients appear to exhibit negligible Reynolds number dependence especially when considering the magnitude of the error at the lowest Reynolds number.

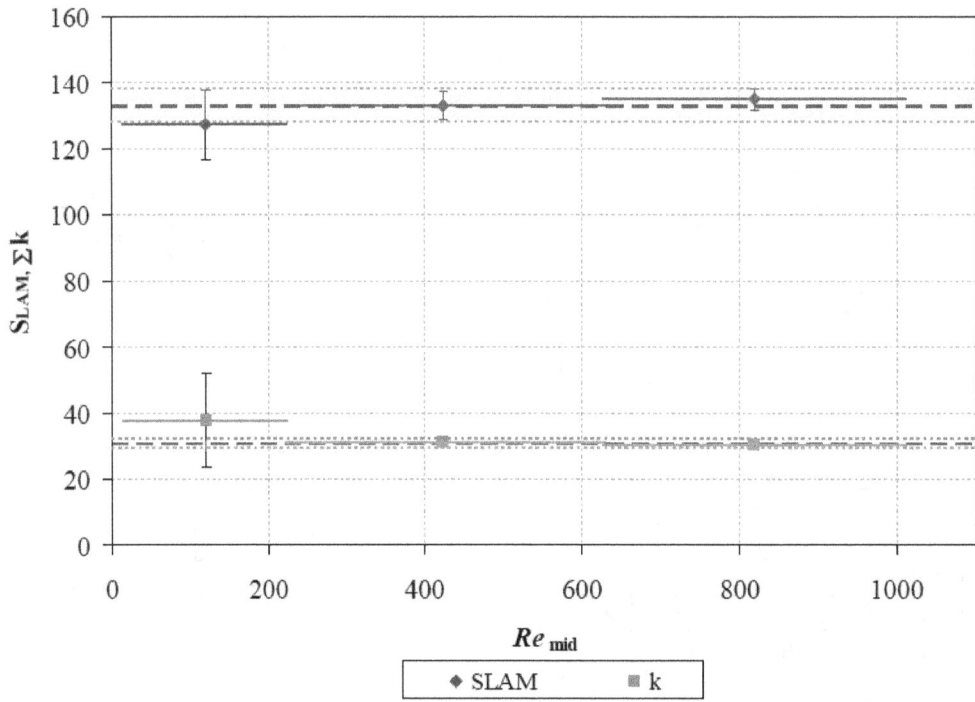

**Figure 3.4   Overall assembly (A–36) hydraulic loss coefficients as a function of Reynolds number for the 217.5 mm storage cell.**

## 3.2   221.8 mm Storage Cell

The 221.8 mm (8.743 in.) size storage cell can be found in the Holtec MPC-24E/EF cask. This size of storage cell is the smallest used in Holtec PWR storage casks.

### 3.2.1   Overall Pressure Drop

#### *3.2.1.1   Overall Pressure Drop Dependence on Reynolds Number with Blocked Guide Tubes*

Figure 3.5 shows the overall assembly pressure drop as a function of Reynolds number in the bundle of the PWR and BWR assemblies. The PWR assembly is contained inside of the mid-sized, 221.8 mm storage cell. This size storage cell represents the smallest commonly found in commercial dry storage casks. The water rods in the BWR and the guide tubes in the PWR are plugged. At all Reynolds numbers, the pressure drop across the PWR is similar to the pressure drop across the BWR assembly. The pressure drop across the BWR assembly is slightly lower than the pressure drop across the PWR assembly at low Reynolds numbers and nearly equal at higher Reynolds numbers.

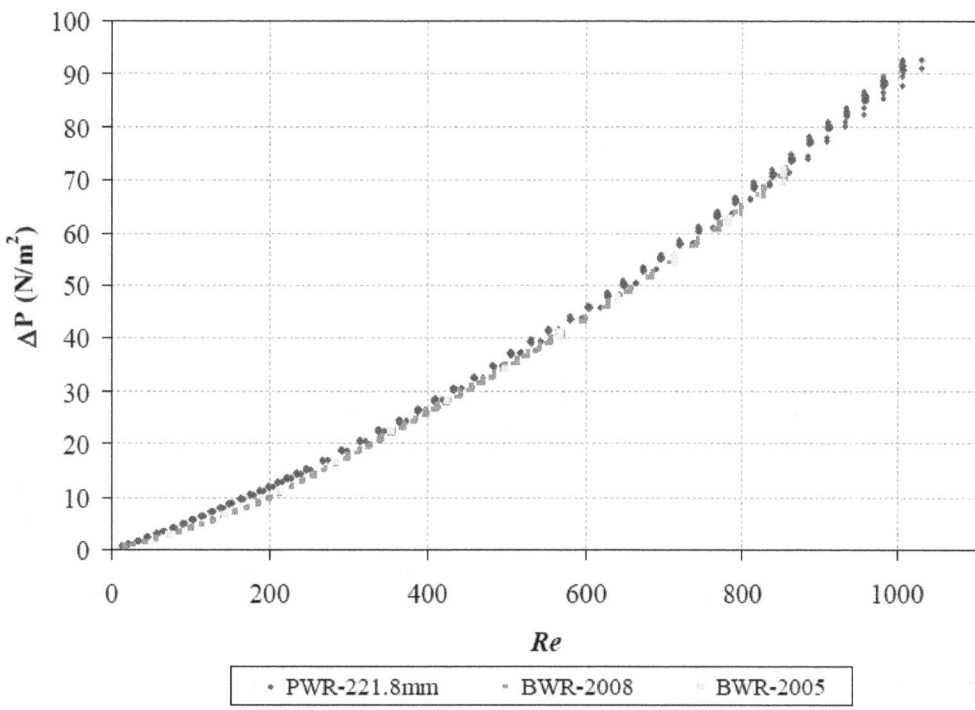

**Figure 3.5** **Pressure drop as a function of Reynolds numbers across the BWR fuel assembly and the PWR fuel assembly in the 221.8 mm storage cell.**

### 3.2.1.2 *Overall Pressure Drop Dependence on Reynolds Number with Unblocked Guide Tubes*

Figure 3.6 shows the overall assembly pressure drop as a function of Reynolds number across the PWR assembly in the 221.8 mm storage cell with blocked and unblocked guide tubes. With the guide tubes open to flow, the flow area of the assembly increases from 0.0275 $m^2$ to 0.0299 $m^2$ but the hydraulic diameter remains unchanged (see Table 2.3). At all Reynolds numbers, the pressure drop across the unblocked PWR is greater than to the pressure drop across the blocked PWR assembly. The increase in flow area provided by the unblocked guide tubes reduces the average flow velocity and Reynolds number for any given volumetric flow rate.

Figure 3.7 presents the same pressure drop data for blocked and unblocked guide tubes but now as a function of total flow rate through the assembly. At any given flow rate, the pressure drop across the assembly with unblocked guide tubes is less than the pressure drop across the assembly with blocked guide tubes. The difference is due to the increased flow through the unblocked guide tubes.

The two data sets in Figure 3.7 were each fit to a quadratic equation with excellent correlation results. The quadratic correlations were solved for flow at a number of pressure drops and differenced to calculate the flow rate through the guide tubes. Figure 3.8 shows the fraction of flow through the guide tubes as a function of total flow through the PWR assembly with unblocked guide tubes. The fraction of flow in the guide tubes drops from about 9% at low flows to about 5% at high flows.

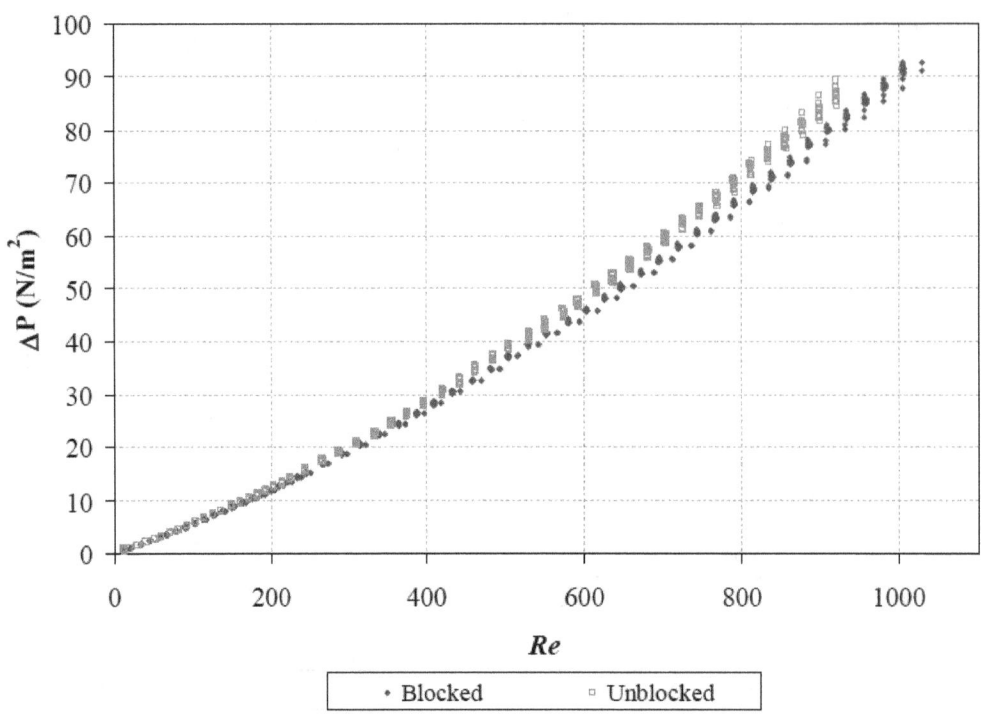

**Figure 3.6** Pressure drop as a function of Reynolds numbers across the PWR fuel assembly in the 221.8 mm storage cell with blocked and unblocked guide tubes.

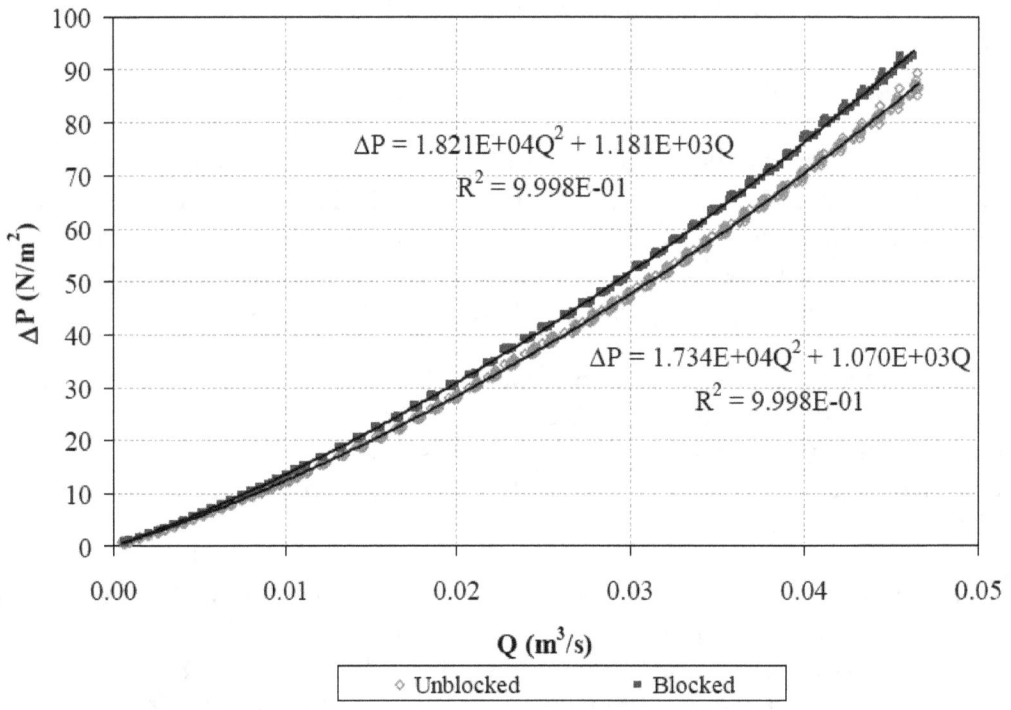

$$\Delta P = 1.821E+04Q^2 + 1.181E+03Q$$
$$R^2 = 9.998E-01$$

$$\Delta P = 1.734E+04Q^2 + 1.070E+03Q$$
$$R^2 = 9.998E-01$$

**Figure 3.7** Pressure drop as a function of volumetric flow rate across the PWR fuel assembly in the 221.8 mm storage cell with blocked and unblocked guide tubes.

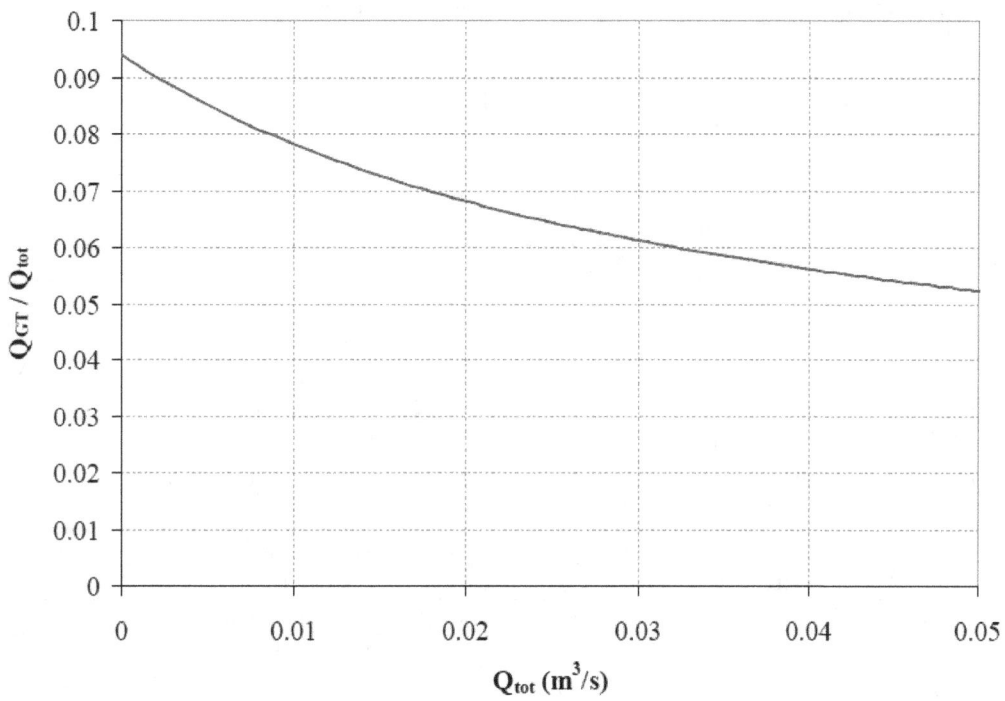

**Figure 3.8** Fraction of flow in guide tubes as a function of total flow through the PWR fuel assembly in the 221.8 mm storage cell with unblocked guide tubes.

## 3.2.2 Hydraulic Loss Coefficients

### 3.2.2.1 Full Flow Range $S_{LAM}$-$\Sigma k$ Analysis with Blocked Guide Tubes

A summary of the $S_{LAM}$ and $\Sigma k$ coefficients for the 221.8 mm storage cell tested with blocked guide tubes is shown in Table 3.3. These values were determined from the full experimental flow rate range of 30 to 2100 slpm, or Reynolds numbers of 10 to 1000, respectively. The reference hydraulic diameter and flow area of the upper section was used to calculate the $S_{LAM}$ and $\Sigma k$ values for any span including the upper section, *e.g.*, A–36. Values of $S_{LAM}$ and $\Sigma k$ for segments 34–36 and 30–34 are again calculated in two ways: 1) using the reference hydraulics of upper bundle ($D_{H, Ref.}$ = 0.0113 m) section and 2) using the reference hydraulics of lower bundle section ($D_{H, Ref.}$ = 0.0116 m). The second set of $S_{LAM}$ and $\Sigma k$ represent the actual local hydraulic conditions. The first set of $S_{LAM}$ and $\Sigma k$ parameters are needed to appropriately integrate the segment results for comparison with the overall assembly (A–36). The directly observed value of $S_{LAM}$ for the overall pressure drop was 109.9, which denotes an error of less than 0.5% from the length-averaged value. The directly measured value of $\Sigma k$ for the overall assembly pressure drop was 27.7, which indicates an error of about 0.7%.

23

**Table 3.3** Full flow range $S_{LAM}$ and $\Sigma k$ coefficients for the PWR assembly in the 221.8 mm storage cell with blocked guide tubes.

| Segment | Description | $D_{H, Ref.}$ (m) | $S_{LAM}$ | $\Sigma k$ | L (m) | $S_{LAM} \cdot (L/L_{tot})$ |
|---------|-------------|-------------------|-----------|------------|-------|------------------------------|
| A–3 | Top Nozzle | 0.0113 | 113.2 | 1.4 | 0.2097 | 5.9 |
| 3–5 | Long Bundle | 0.0113 | 65.6 | 1.2 | 0.4509 | 7.3 |
| 5–6 | Spacer | 0.0113 | 357.6 | 1.9 | 0.0667 | 5.9 |
| 6–7 | Short Bundle | 0.0113 | 57.7 | 0.9 | 0.1990 | 2.8 |
| 7–8 | IFM | 0.0113 | 231.2 | 1.1 | 0.0545 | 3.1 |
| 8–9 | Short Bundle | 0.0113 | 56.0 | 0.5 | 0.2021 | 2.8 |
| 9–10 | Spacer | 0.0113 | 364.6 | 1.5 | 0.0656 | 5.9 |
| 10–11 | Short Bundle | 0.0113 | 68.3 | 0.9 | 0.1979 | 3.3 |
| 11–12 | IFM | 0.0113 | 237.2 | 1.1 | 0.0545 | 3.2 |
| 12–13 | Short Bundle | 0.0113 | 50.7 | 0.5 | 0.2043 | 2.6 |
| 13–14 | Spacer | 0.0113 | 394.9 | 1.9 | 0.0661 | 6.5 |
| 14–15 | Short Bundle | 0.0113 | 57.4 | 0.7 | 0.1995 | 2.8 |
| 15–16 | IFM | 0.0113 | 276.7 | 1.4 | 0.0534 | 3.7 |
| 16–17 | Short Bundle | 0.0113 | 47.5 | 0.3 | 0.2027 | 2.4 |
| 17–18 | Spacer | 0.0113 | 383.5 | 1.8 | 0.0646 | 6.1 |
| 18–22 | Long Bundle | 0.0113 | 62.4 | 1.2 | 0.4604 | 7.1 |
| 22–23 | Spacer | 0.0113 | 366.7 | 1.8 | 0.0646 | 5.8 |
| 23–29 | Long Bundle | 0.0113 | 65.4 | 1.5 | 0.4710 | 7.6 |
| 29–30 | Spacer | 0.0113 | 314.5 | 1.3 | 0.0667 | 5.2 |
| 30–34 | Long Bundle | 0.0113 | 73.0 | 1.6 | 0.5244 | 9.5 |
| 30–34 | Long Bundle | 0.0116 | 77.9 | 1.7[†] | -- | -- |
| 34–36 | Bottom Nozzle | 0.0113 | 238.7 | 3.4 | 0.1688 | 10.0 |
| 34–36 | Bottom Nozzle | 0.0116 | 255.0 | 3.5[†] | -- | -- |
| Summation | Overall (equiv.) | 0.0113 | -- | 27.9 | 4.0472 | 109.4 |
| A–36 | Overall (meas.) | 0.0113 | 109.9 | 27.7 | 4.0472 | -- |

[†] These form losses are not included in the sum total.

### 3.2.2.2 Full Flow Range $S_{LAM}$-$\Sigma k$ Analysis with Unblocked Guide Tubes

A summary of the $S_{LAM}$ and $\Sigma k$ coefficients for the 221.8 mm storage cell testing with unblocked guide tubes is shown in Table 3.4. These values were determined from the full experimental flow rate range of 30 to 2100 slpm, or Reynolds numbers of 10 to 1000, respectively. The reference hydraulic diameter and flow area of the upper section was used to calculate the $S_{LAM}$ and $\Sigma k$ values for any span including the upper section, *e.g.*, A–36. The increased flow area and wetted perimeter for the inside of the guide tubes is included in these calculations. As in the previous sections, values of $S_{LAM}$ and $\Sigma k$ for segments 34–36 and 30–34 are again calculated in two ways: 1) using the reference hydraulics of upper bundle ($D_{H, Ref.}$ = 0.0113 m) section and 2) using the reference hydraulics of lower bundle section ($D_{H, Ref.}$ = 0.0116 m). The second set of $S_{LAM}$ and $\Sigma k$ represent the actual local hydraulic conditions. The first set of $S_{LAM}$ and $\Sigma k$ parameters are needed to appropriately integrate the segment results for comparison with the overall assembly (A–36). Since the lower guide tube section does not carry flow, the lower bundle hydraulics are the same in the blocked and unblocked cases. The directly observed value of $S_{LAM}$ for the overall pressure drop was 108.6, which denotes an error of less than 1.3% from

24

the length-averaged value. The directly measured value of $\Sigma k$ for the overall assembly pressure drop was 31.7, which indicates an error of about 0.6%.

At a value of 108.6, the $S_{LAM}$ for the unblocked case is not significantly lower than $S_{LAM} = 109.9$ for the blocked guide tube case. However, the difference in the form loss parameter, $\Sigma k$, is more significant. For the unblocked case, the form losses are 31.7. This represents a substantial increase from the blocked case of 27.7 and is well above the experimental uncertainty of $u_k = 1.4$.

Table 3.4    Full flow range $S_{LAM}$ and $\Sigma k$ coefficients for the PWR assembly in the 221.8 mm storage cell with unblocked guide tubes.

| Segment | Description | $D_{H, Ref.}$ (m) | $S_{LAM}$ | $\Sigma k$ | L (m) | $S_{LAM} \cdot (L/L_{tot})$ |
|---|---|---|---|---|---|---|
| A–3 | Top Nozzle | 0.0113 | 116.7 | 1.5 | 0.2097 | 6.0 |
| 3–5 | Long Bundle | 0.0113 | 60.6 | 1.6 | 0.4509 | 6.8 |
| 5–6 | Spacer | 0.0113 | 339.8 | 2.2 | 0.0667 | 5.6 |
| 6–7 | Short Bundle | 0.0113 | 58.9 | 0.9 | 0.1990 | 2.9 |
| 7–8 | IFM | 0.0113 | 235.7 | 1.2 | 0.0545 | 3.2 |
| 8–9 | Short Bundle | 0.0113 | 52.0 | 0.6 | 0.2021 | 2.6 |
| 9–10 | Spacer | 0.0113 | 335.6 | 1.9 | 0.0656 | 5.4 |
| 10–11 | Short Bundle | 0.0113 | 59.9 | 1.0 | 0.1979 | 2.9 |
| 11–12 | IFM | 0.0113 | 239.9 | 1.2 | 0.0545 | 3.2 |
| 12–13 | Short Bundle | 0.0113 | 49.4 | 0.6 | 0.2043 | 2.5 |
| 13–14 | Spacer | 0.0113 | 389.1 | 2.1 | 0.0661 | 6.4 |
| 14–15 | Short Bundle | 0.0113 | 56.9 | 0.8 | 0.1995 | 2.8 |
| 15–16 | IFM | 0.0113 | 263.8 | 1.6 | 0.0534 | 3.5 |
| 16–17 | Short Bundle | 0.0113 | 46.6 | 0.4 | 0.2027 | 2.3 |
| 17–18 | Spacer | 0.0113 | 358.1 | 2.0 | 0.0646 | 5.7 |
| 18–22 | Long Bundle | 0.0113 | 62.3 | 1.5 | 0.4604 | 7.1 |
| 22–23 | Spacer | 0.0113 | 364.0 | 1.9 | 0.0646 | 5.8 |
| 23–29 | Long Bundle | 0.0113 | 61.9 | 1.9 | 0.4710 | 7.2 |
| 29–30 | Spacer | 0.0113 | 293.5 | 1.5 | 0.0667 | 4.8 |
| 30–34 | Long Bundle | 0.0113 | 78.2 | 1.7 | 0.5244 | 10.1 |
| 30–34 | Long Bundle | 0.0116 | 76.5 | 1.5† | -- | -- |
| 34–36 | Bottom Nozzle | 0.0113 | 248.5 | 4.0 | 0.1688 | 10.4 |
| 34–36 | Bottom Nozzle | 0.0116 | 243.2 | 3.5† | -- | -- |
| Summation | Overall (equiv.) | 0.0113 | -- | 31.9 | 4.0472 | 107.3 |
| A–36 | Overall (meas.) | 0.0113 | 108.6 | 31.7 | 4.0472 | -- |

† These form losses are not included in the sum total.

### 3.2.2.3   Partitioned Flow Range $S_{LAM}$-$\Sigma k$ Analysis with Blocked Guide Tubes

Figure 3.9 shows the $S_{LAM}$ and $\Sigma k$ for the three-partition analysis of data for the 221.8 mm cell. The points represent the $S_{LAM}$ and $\Sigma k$ for the corresponding average Reynolds number of the data used in the analysis. The horizontal bars represent the range of Reynolds numbers over which the hydraulic loss coefficients were calculated. The vertical bars represent the error for the $S_{LAM}$ and $\Sigma k$ calculated by the method detailed in Appendix B. The error is large for small Reynolds numbers and decreases as the Reynolds number increases. The dashed black lines represent the respective overall $S_{LAM}$ and $\Sigma k$ values reported in Table 3.3. The yellow shading about the black dashed lines represents the experimental error associated with the overall $S_{LAM}$ and $\Sigma k$ values. The error range for all the partitioned data overlaps with the error range of the overall $S_{LAM}$ and $\Sigma k$. The hydraulic loss coefficients appear to exhibit negligible Reynolds number dependence.

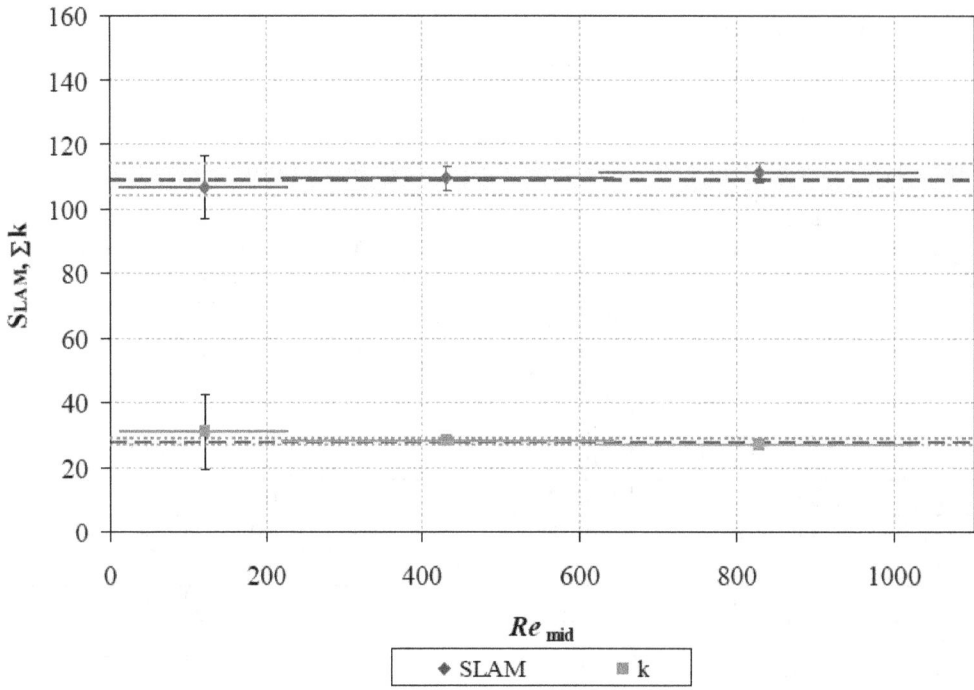

**Figure 3.9**   **Overall assembly (A–36) hydraulic loss coefficients as a function of Reynolds number for the 221.8 mm storage cell and blocked guide tubes.**

### 3.2.2.4   *Partitioned Flow Range S$_{LAM}$-Σk Analysis with Unblocked Guide Tubes*

Figure 3.10 shows the $S_{LAM}$ and $\Sigma k$ for the three-partition analysis of data for the 221.8 mm cell with unblocked guide tubes. The points represent the $S_{LAM}$ and $\Sigma k$ for the corresponding average Reynolds number of the data used in the analysis. The horizontal bars represent the range of Reynolds numbers over which the hydraulic loss coefficients were calculated. The vertical bars represent the error for the $S_{LAM}$ and $\Sigma k$ calculated by the method detailed in Appendix B. The error is large for small Reynolds numbers and decreases as the Reynolds number increases. The dashed black lines represent the respective overall $S_{LAM}$ and $\Sigma k$ values reported in Table 3.4. The yellow shading about the black dashed lines represents the experimental error associated with the overall $S_{LAM}$ and $\Sigma k$ values. The error range for all the partitioned data overlaps with the error range of the overall $S_{LAM}$ and $\Sigma k$. The hydraulic loss coefficients appear to exhibit negligible Reynolds number dependence.

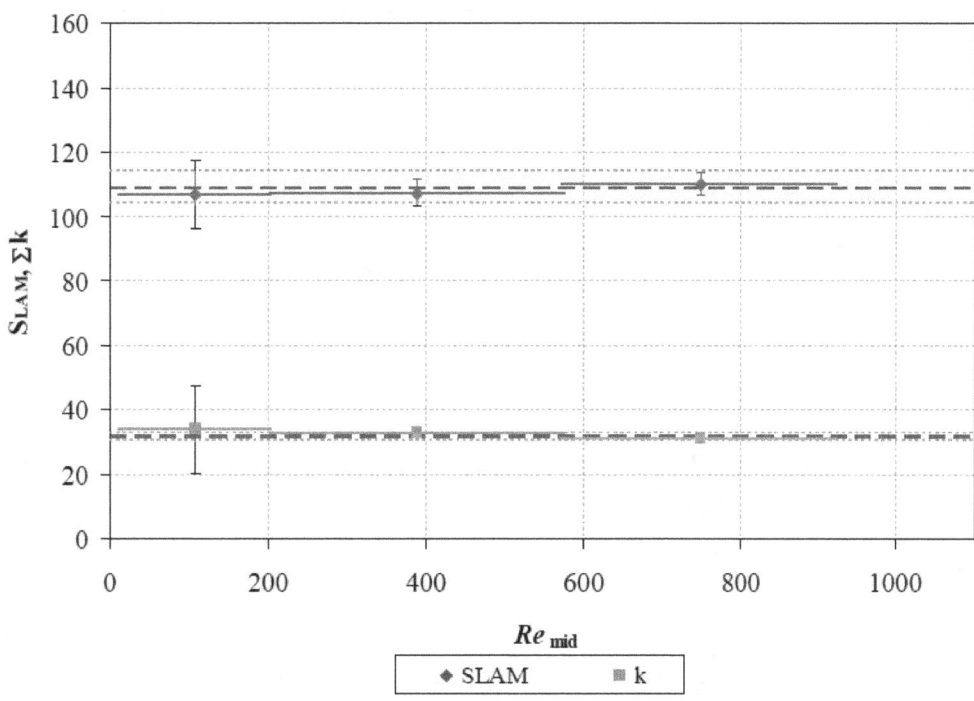

**Figure 3.10 Overall assembly (A–36) hydraulic loss coefficients as a function of Reynolds number for the 221.8 mm storage cell and unblocked guide tubes.**

## 3.3   226.6 mm Storage Cell

The 226.6 mm (8.920 in.) size storage cell can be found in the Holtec Holtec MPC-24 cask. This size storage cell is among the largest used in Holtec dry casks. All of the 24 cells in the MPC-24 cask are this size. There are a few large cells in other Holtec casks, but they are not significant in number. The MPC-32 casks has two (of 32) cells that are 227.1 mm and the MPC-24E/EF has four (of 24) cells that are 229.9 mm.

### 3.3.1   Overall Pressure Drop

#### *3.3.1.1   Overall Pressure Drop Dependence on Reynolds Number*

Figure 3.11 shows the overall assembly pressure drop as a function of Reynolds number in the bundle of the PWR and BWR assemblies. The PWR assembly is contained inside of the largest sized, 226.6 mm storage cell. This size storage cell represents the largest commonly found in commercial dry storage casks. The water rods in the BWR and the guide tubes in the PWR are blocked. At all Reynolds numbers, the pressure drop across the PWR is significantly less than the pressure drop across the BWR assembly.

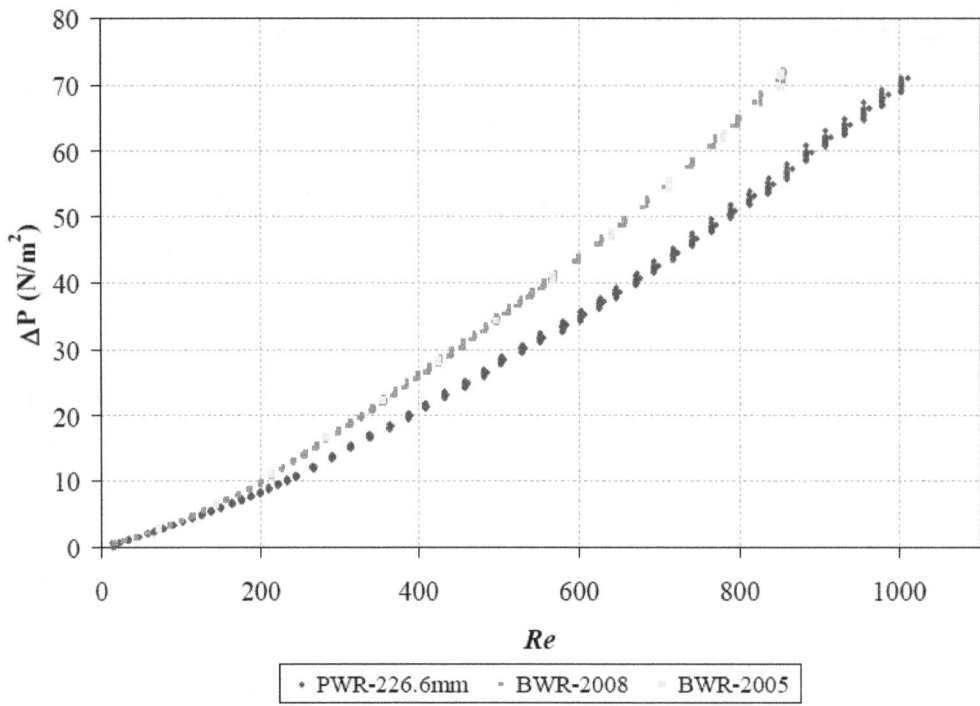

**Figure 3.11  Pressure drop as a function of Reynolds numbers across the BWR fuel assembly and the PWR fuel assembly in the 226.6 mm storage cell.**

### 3.3.2   Hydraulic Loss Coefficients

#### 3.3.2.1   *Full Flow Range $S_{LAM}$-$\Sigma k$ Analysis*

A summary of the $S_{LAM}$ and $\Sigma k$ coefficients for the 226.6 mm storage cell testing is shown in Table 3.5. These values were determined from the full experimental flow rate range of 30 to 2100 slpm, or Reynolds numbers of 10 to 1000, respectively. The reference hydraulic diameter and flow area of the upper section was used to calculate the $S_{LAM}$ and $\Sigma k$ values for any span including the upper section, *e.g.*, A–36. Values of $S_{LAM}$ and $\Sigma k$ for segments 34–36 and 30–34 are calculated in two ways:  1) using the reference hydraulics of upper bundle ($D_{H, Ref.}$ = 0.0121 m) section and 2) using the reference hydraulics of lower bundle section ($D_{H, Ref.}$ = 0.0124 m). The second set of $S_{LAM}$ and $\Sigma k$ represent the actual local hydraulic conditions. The first set of $S_{LAM}$ and $\Sigma k$ parameters are needed to appropriately integrate the segment results for comparison with the overall (A–36). The directly observed value of $S_{LAM}$ for the overall pressure drop was 98.5, which denotes an error of about 1% from the length-averaged value. The directly measured value of $\Sigma k$ for the overall assembly pressure drop was 27.4, which indicates an error of about 1.1%.

28

**Table 3.5** Full flow range $S_{LAM}$ and $\Sigma k$ coefficients for the PWR assembly in the 226.6 mm storage cell.

| Segment | Description | $D_{H, Ref.}$ (m) | $S_{LAM}$ | $\Sigma k$ | L (m) | $S_{LAM} \cdot (L/L_{tot})$ |
|---|---|---|---|---|---|---|
| A–3 | Top Nozzle | 0.0121 | 89.9 | 1.4 | 0.2097 | 4.7 |
| 3–5 | Long Bundle | 0.0121 | 65.7 | 1.4 | 0.4509 | 7.3 |
| 5–6 | Spacer | 0.0121 | 303.4 | 1.6 | 0.0667 | 5.0 |
| 6–7 | Short Bundle | 0.0121 | 52 | 0.7 | 0.1990 | 2.6 |
| 7–8 | IFM | 0.0121 | 227.7 | 1.0 | 0.0545 | 3.1 |
| 8–9 | Short Bundle | 0.0121 | 51.7 | 0.5 | 0.2021 | 2.6 |
| 9–10 | Spacer | 0.0121 | 269.2 | 1.6 | 0.0656 | 4.4 |
| 10–11 | Short Bundle | 0.0121 | 57.3 | 1.1 | 0.1979 | 2.8 |
| 11–12 | IFM | 0.0121 | 194.3 | 1.0 | 0.0545 | 2.6 |
| 12–13 | Short Bundle | 0.0121 | 53.3 | 0.6 | 0.2043 | 2.7 |
| 13–14 | Spacer | 0.0121 | 302.1 | 1.8 | 0.0661 | 4.9 |
| 14–15 | Short Bundle | 0.0121 | 53.9 | 0.8 | 0.1995 | 2.7 |
| 15–16 | IFM | 0.0121 | 223.1 | 1.2 | 0.0534 | 2.9 |
| 16–17 | Short Bundle | 0.0121 | 45.7 | 0.4 | 0.2027 | 2.3 |
| 17–18 | Spacer | 0.0121 | 341.9 | 1.8 | 0.0646 | 5.5 |
| 18–22 | Long Bundle | 0.0121 | 56 | 1.2 | 0.4604 | 6.4 |
| 22–23 | Spacer | 0.0121 | 322.7 | 1.6 | 0.0646 | 5.1 |
| 23–29 | Long Bundle | 0.0121 | 60 | 1.7 | 0.4710 | 7.0 |
| 29–30 | Spacer | 0.0121 | 296.7 | 1.6 | 0.0667 | 4.9 |
| 30–34 | Long Bundle | 0.0121 | 68.1 | 1.5 | 0.5244 | 8.8 |
| 30–34 | Long Bundle | 0.0124 | 72.4 | 1.5[†] | -- | -- |
| 34–36 | Bottom Nozzle | 0.0121 | 223.5 | 3.3 | 0.1688 | 9.3 |
| 34–36 | Bottom Nozzle | 0.0124 | 237.8 | 3.4[†] | -- | -- |
| Summation | Overall (equiv.) | 0.0121 | -- | 27.7 | 4.0472 | 97.5 |
| A–36 | Overall (meas.) | 0.0121 | 98.5 | 27.4 | 4.0472 | -- |

[†] These form losses are not included in the sum total.

### 3.3.2.2 Partitioned Flow Range $S_{LAM}$-$\Sigma k$ Analysis

Figure 3.12 shows the $S_{LAM}$ and $\Sigma k$ for the three-partition analysis of data for the 226.6 mm cell. The points represent the $S_{LAM}$ and $\Sigma k$ for the corresponding average Reynolds number of the data used in the analysis. The horizontal bars represent the range of Reynolds numbers over which the hydraulic loss coefficients were calculated. The vertical bars represent the error for the $S_{LAM}$ and $\Sigma k$ calculated by the method detailed in Appendix B. The error is large for small Reynolds numbers and decreases as the Reynolds number increases. The dashed black lines represent the respective overall $S_{LAM}$ and $\Sigma k$ values reported in Table 3.5. The yellow shading about the black dashed lines represents the experimental error associated with the overall $S_{LAM}$ and $\Sigma k$ values. The error range for almost all the partitioned data overlaps with the error range of the overall $S_{LAM}$ and $\Sigma k$. The hydraulic loss coefficients appear to exhibit a slight Reynolds number dependence but the magnitude of the error at the lowest Reynolds numbers casts doubt on this interpretation.

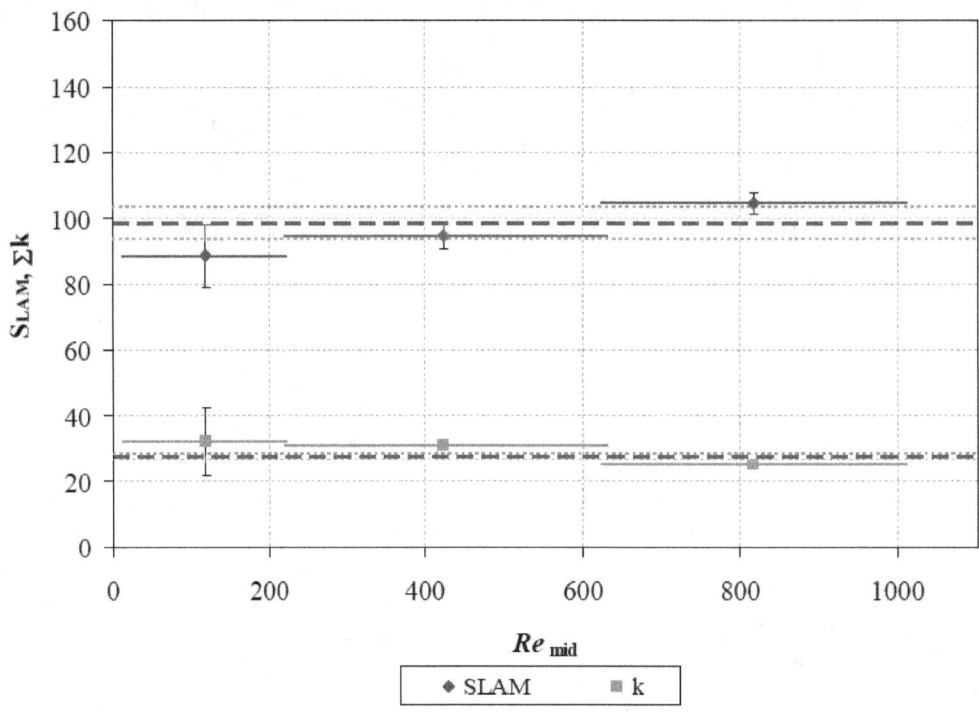

**Figure 3.12 Overall assembly (A–36) hydraulic loss coefficients as a function of Reynolds number for the 226.6 mm storage cell.**

## 3.4 Boiling Water Reactor Assembly

### 3.4.1 Hydraulic Loss Coefficients

#### 3.4.1.1 Full Flow Range $S_{LAM}$-$\Sigma k$ Analysis with Blocked Water Rods

A summary of the $S_{LAM}$ and $\Sigma k$ coefficients for the BWR assembly with blocked water rods is shown in Table 3.6. These measurements repeat those conducted in 2005 at SNL but include a more refined flow control capability. The reference hydraulic loss coefficients calculated in the present investigations are within the experimental error from those documented in SAND2007-2270 (2005 – $S_{LAM}$ = 106, $\Sigma k$ = 37).[1] These values were determined from the full experimental flow rate range of 15 to 600 slpm, or Reynolds numbers of 20 to 850, respectively. The hydraulic diameter and flow area of the upper and lower sections of the BWR assembly are listed in Table 2.3. The hydraulic diameter and flow area of the lower fully populated bundle section were used to calculate the $S_{LAM}$ and $\Sigma k$ values for any span including the lower section, *e.g.*, 1–B. Values of $S_{LAM}$ and $\Sigma k$ for segments in the upper section are calculated in two ways: 1) using the reference hydraulics of lower section ($D_{H, Ref}$ = 0.0119 m) section and 2) using the reference hydraulics of upper section ($D_{H, Ref}$ = 0.0141 m). The second set of $S_{LAM}$ and $\Sigma k$ represent the actual local hydraulic conditions. The first set of $S_{LAM}$ and $\Sigma k$ parameters are needed to appropriately integrate the segment results for comparison with the overall assembly (1–B). The directly observed value of $S_{LAM}$ for the overall pressure drop was 104.2, which denotes an error of less than 1% from the length-averaged value. The directly measured value of $\Sigma k$ for the overall assembly pressure drop was 37.8, which also indicates an error of less than 1%.

30

**Table 3.6    Full flow range $S_{LAM}$ and $\Sigma k$ coefficients for the BWR assembly with blocked water rods.**

| Segment | Description | $D_{H, Ref.}$ (m) | $S_{LAM}$ | $\Sigma k$ | L (m) | $S_{LAM} \cdot (L/L_{tot})$ |
|---|---|---|---|---|---|---|
| 1–2 | Top Tie | 0.0119 | 44.4 | 0.4 | 0.040 | 0.4 |
| 1–2 | Top Tie | 0.0141 | 66.7 | 0.4[†] | -- | -- |
| 2–3 | Long Bundle | 0.0119 | 34.8 | 0.3 | 0.411 | 3.4 |
| 2–3 | Long Bundle | 0.0141 | 52.3 | 0.3[†] | -- | -- |
| 3–4 | Spacer | 0.0119 | 313.3 | 2.6 | 0.044 | 3.3 |
| 3–4 | Spacer | 0.0141 | 471.0 | 3.1[†] | -- | -- |
| 4–6 | Long Bundle | 0.0119 | 38.5 | 0.9 | 0.468 | 4.3 |
| 4–6 | Long Bundle | 0.0141 | 57.8 | 1.0[†] | -- | -- |
| 6–7 | Spacer | 0.0119 | 298.6 | 2.6 | 0.042 | 3.0 |
| 6–7 | Spacer | 0.0141 | 448.9 | 3.0[†] | -- | -- |
| 7–8 | Long Bundle | 0.0119 | 48.0 | 0.5 | 0.469 | 5.4 |
| 7–8 | Long Bundle | 0.0141 | 72.2 | 0.5[†] | -- | -- |
| 8–9 | Spacer | 0.0119 | 733.6 | 3.4 | 0.044 | 7.8 |
| 9–10 | 2×Long Bundles + Spacer | 0.0119 | 105.4 | 5.0 | 0.980 | 24.7 |
| 10–11 | Spacer | 0.0119 | 707.7 | 3.4 | 0.044 | 7.5 |
| 11–13 | Long Bundle | 0.0119 | 77.2 | 0.9 | 0.467 | 8.6 |
| 13–14 | Spacer | 0.0119 | 720.3 | 3.3 | 0.044 | 7.6 |
| 14–15 | Long Bundle | 0.0119 | 75.1 | 0.8 | 0.468 | 8.4 |
| 15–16 | Spacer | 0.0119 | 739.1 | 3.3 | 0.043 | 7.5 |
| 16–17 | Long Bundle | 0.0119 | 73.0 | 0.7 | 0.470 | 8.2 |
| 17–B | Bottom Tie | 0.0119 | 124.5 | 10.0 | 0.127 | 2.3 |
| Summation | Overall (equiv.) | 0.0119 | -- | 38.1 | 4.190 | 103.4 |
| 1–B | Overall (meas.) | 0.0119 | 104.2 | 37.8 | 4.190 | -- |

### 3.4.1.2  *Partitioned Flow Range $S_{LAM}$-$\Sigma k$ Analysis with Blocked Water Rods*

Figure 3.13 shows the $S_{LAM}$ and $\Sigma k$ for the three-partition analysis of data for the BWR assembly. The horizontal bars represent the range of Reynolds numbers over which the hydraulic loss coefficients were calculated. The vertical bars represent the error for the $S_{LAM}$ and $\Sigma k$ calculated by the method detailed in Appendix B. The error is large for small Reynolds numbers and decreases as the Reynolds number increases. The dashed black lines represent the respective overall $S_{LAM}$ and $\Sigma k$ values reported in Table 3.6. The yellow shading about the black dashed lines represents the experimental error associated with the overall $S_{LAM}$ and $\Sigma k$ values. The hydraulic loss coefficients appear to exhibit a Reynolds number dependence. The $S_{LAM}$ increases from 94 to 109 at $Re_{mid} = 75$ to 770, respectively. Similarly, the $\Sigma k$ decreases from 69 to 35 at $Re_{mid} = 75$ to 770, respectively.

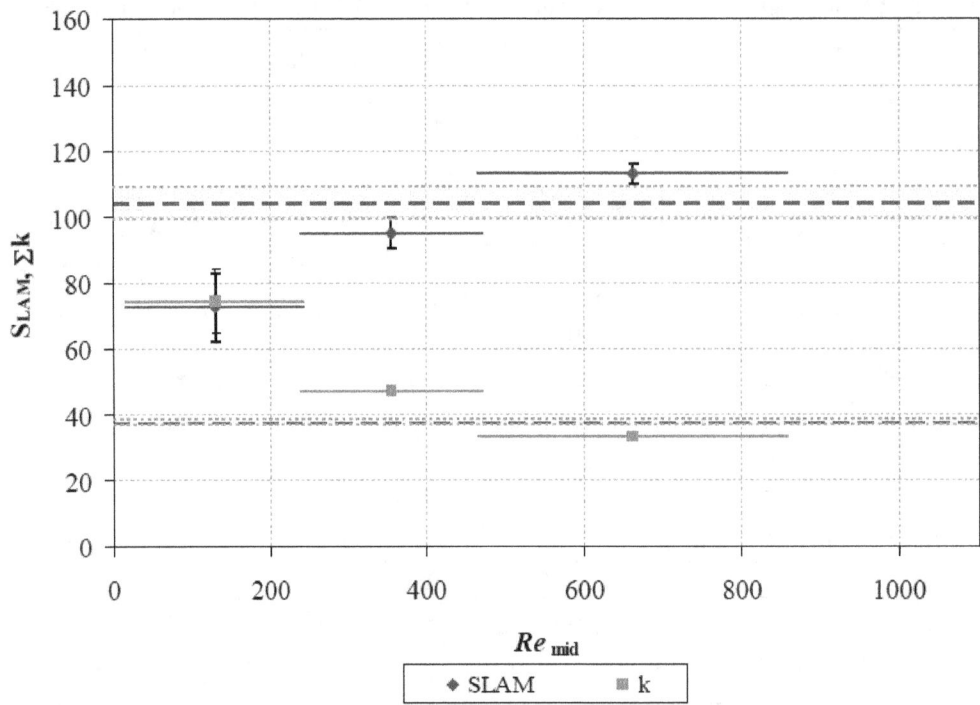

**Figure 3.13 Overall assembly (1–B) hydraulic loss coefficients as a function of Reynolds number for the BWR with blocked water rods.**

## 3.5    PWR Storage Cell Size Dependence

Three different sized PWR storage cells were tested to determine the effect of the storage cell size on the hydraulic viscous and form loss parameters.  The smallest storage cell tested was 217.5 mm and is constrained by the size of the spacer, which is 214 mm.  The other two storage cells tested were 221.8 mm and 226.6 mm, which span the sizes of the most common storage cells.  The largest storage cell found in the three dry casks considered is approximately 230 mm (see Table 2.2).

### 3.5.1    Full Flow Range

The geometry of the annular flow path is fundamentally different in each of the storage cells.  The hydraulic diameter of the storage cell increases as the flow area of the annular region increases.  This change in annular flow areas between different storage cells affects the distribution of flow between the bundle and annular regions.  The hydraulic loss parameters are also affected.    Figure 3.14 shows the dependence of the $S_{LAM}$ and $\Sigma k$ on the storage cell hydraulic diameter based on data from the full range of laminar flows ($Re = 10$ to $1000$).

The full-flow-range $S_{LAM}$ drops from 134.0 at $D_{H, Ref.} = 0.0105$ m (217.5 mm cell) to 110.7 at $D_{H, Ref.} = 0.0113$ m (221.8 mm cell) to 98.6 at $D_{H, Ref.} = 0.0121$ m (226.6 mm cell).  An empirical power law correlation was developed to aid in assigning hydraulic parameters to storage cell sizes not tested.  The validity of this correlation is limited to 17×17 PWR fuel.  In the limit as the cells size increases, the $S_{LAM}$ asymptotically approaches the value of 57 for a square duct [Kays and Crawford, Fig 6-4, p. 63].[5]  A power law correlation was chosen because it could be forced to approach this limiting value.

32

The full-flow-range form loss coefficient, $\Sigma k$, shows less dependence on cell size. For $D_{H, Ref.} = 0.0105$ m (217.5 mm cell), $\Sigma k$ is 30.9 and drops to 28.0 at $D_{H, Ref.} = 0.0113$ m (221.8 mm cell) and 27.8 at $D_{H, Ref.} = 0.0121$ m (226.6 mm cell). A power law correlation was also used to fit the form loss data.

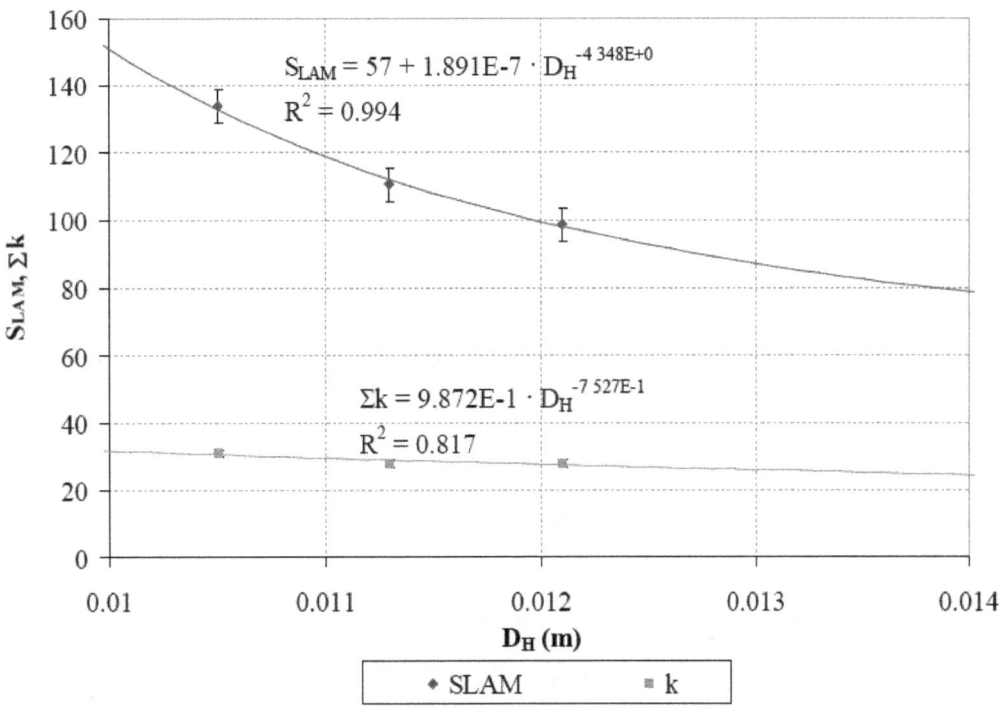

Figure 3.14 **Dependence of overall assembly (A–36) $S_{LAM}$ and $\Sigma k$ on the hydraulic diameter for full data range ($Re = 10$ to $1000$). The validity of this correlation is limited to 17×17 PWR fuel.**

This page intentionally blank

# 4   VELOCITY PROFILE RESULTS

Velocity profiles were measured across the bundle for the three pool cells included in this study. These profiles are valuable in estimating the flow partition between the bundle and annular regions within the assembly. These measurements also indicated a redistribution of flow after spacers and intermediate flow mixers (IFMs) at higher flow rates, suggesting significant wake effects. The wake disturbances in the flow as shown in Section 4.1.3 were generally not apparent in the mid-bundle measurements, which may suggest that the flow has reestablished a fully developed condition.

Figure 4.1 shows the definition of interior and annular cells within the assembly. An interior cell is defined as the interstitial space formed between four fuel rods with an area of $A_{int}$. An interior cell formed by the interstitial space formed between three fuel rods and a guide tube has an area defined as $A_{int\text{-}GT}$. Similarly, an annular cell is the interstitial space formed by two fuel rods and the storage cell wall with an area of $A_{ann,\, cell}$. The distance $L_{ann}$ is the length between the inner storage cell wall and the centerlines of the outermost rod bank. The dashed lines show a portion of the traverses captured by the LDA.

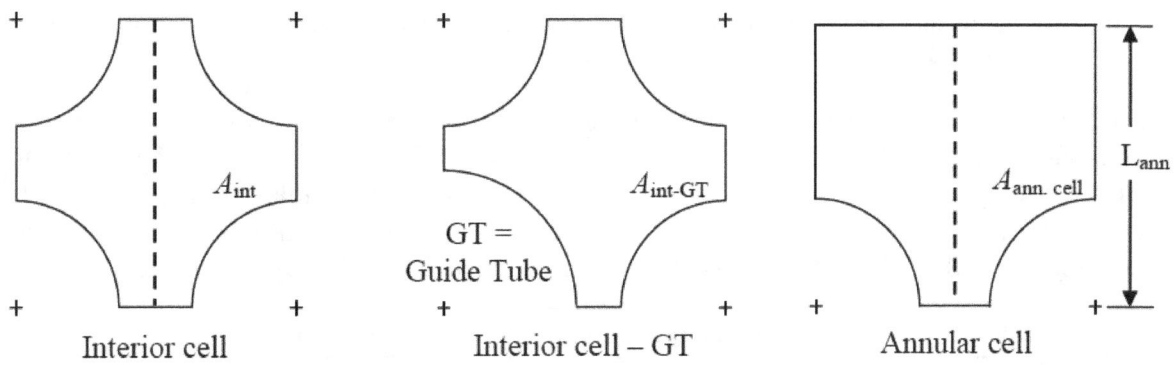

**Figure 4.1   Definition of interior and annular cells for determining the partitioning of flow inside the assembly.**

Table 4.1 shows the area and number of the two types of interior cells. The sum of these cells constitutes the area defined as the bundle in the following analyses. The definition of the bundle and annulus are shown schematically in Figure 4.2. This figure also shows the locations of the LDA traverses. The bundle is taken as the portion of the assembly inside the planes formed by the centerlines of the outer rods. The annular region is the portion of the assembly from this boundary to the inner cell walls.

**Table 4.1   Bundle cell values of area.**

|  | Interior | Interior - GT |
|---|---|---|
| Area (m$^2$) | $8.75\times10^{-5}$ | $7.61\times10^{-5}$ |
| Quantity | 156 | 100 |
| $A_{bundle}$ (m$^2$) | 0.02126 | |

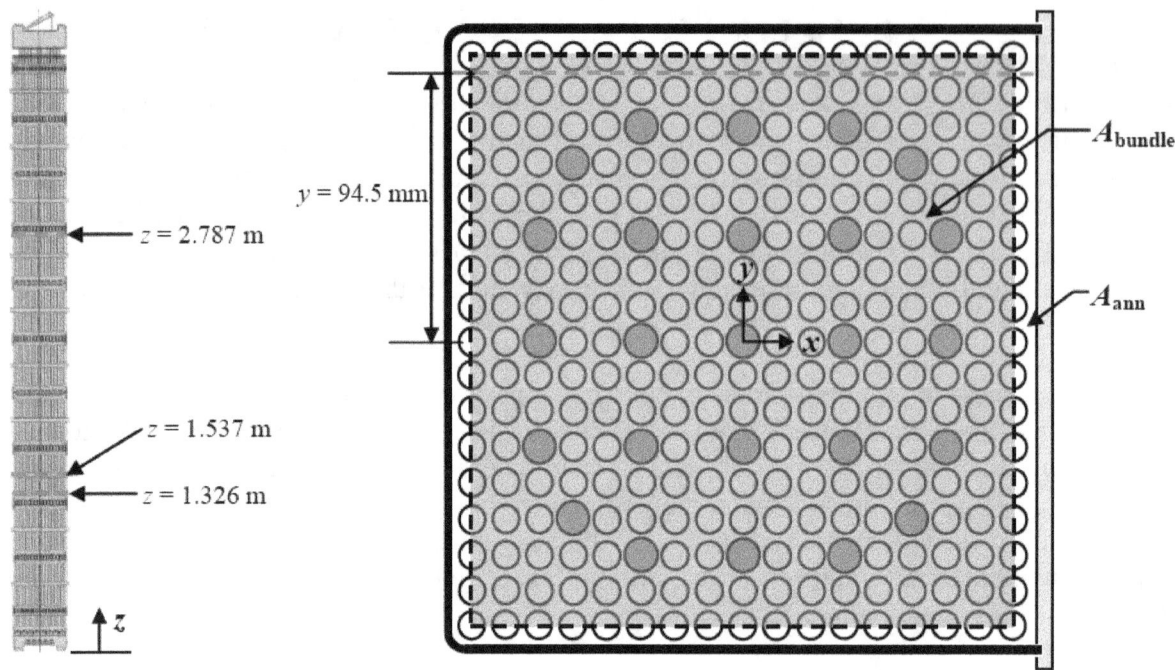

**Figure 4.2** Schematic showing the definition of the bundle (shaded) and annular (white) flow paths in the PWR assembly.

Table 4.2 gives the annular values for the different storage cells used in this study. The annular area to total area ratio are 0.170, 0.227, and 0.282 for the 217.5, 221.8, and 226.6 mm storage cells, respectively. The flow partition between the bundle and annular regions is strongly influenced by this ratio. As the annular area ratio is increased, the amount of flow in this region increases significantly. This effect is discussed in more detail in the following sections.

**Table 4.2** Annular and assembly hydraulic values for the different storage cells.

| Storage Cell (mm) | $L_{ann}$ (mm) | $A_{ann, cell}$ (m$^2$) | $A_{ann}$ (m$^2$) | $A_{tot}$ (m$^2$) |
|---|---|---|---|---|
| 217.5 | 7.96 | $6.47\times10^{-5}$ | 0.00434 | 0.02560 |
| 221.8 | 10.11 | $9.18\times10^{-5}$ | 0.00624 | 0.02750 |
| 226.6 | 12.51 | $1.22\times10^{-4}$ | 0.00834 | 0.02960 |

The ratio of the integrated average velocity along the lines shown in Figure 4.1 to the area integrated average velocity is of particular importance in interpreting these results. Since only the line average is available from the LDA measurements, a limited number of computational fluid dynamics (CFD) runs were conducted to determine this ratio. A PWR bundle section with the same geometry was simulated in FLUENT with the laminar solution model to calculate this ratio as summarized in Table 4.3. The area to line ratios of both the interior and annular cells were averaged for the two Reynolds numbers in each storage cell for the LDA analysis. The resulting average ratios are 0.741 (interior) and 0.961 (annular) for the 217.5 mm cell, 0.739 (interior) and 0.903 (annular) for the 221.8 mm cell, and 0.732 (interior) and 0.893 (annular) for the 226.6 mm cell.

**Table 4.3    Velocity values from CFD simulations of the PWR fuel assembly.**

| | Cell Size (mm) | | | | | |
|---|---|---|---|---|---|---|
| | **217.5** | | **221.8** | | **226.6** | |
| $Re$ | 100 | 900 | 100 | 900 | 100 | 900 |
| $W_{inlet}$ (m/s) | 0.1690 | 1.6000 | 0.1570 | 1.4900 | 0.1460 | 1.3870 |
| Interior: $W_{area,\,avg}$ | 0.1904 | 1.7398 | 0.1561 | 1.5022 | 0.1136 | 1.2889 |
| Interior: $W_{line,\,avg}$ | 0.2570 | 2.3458 | 0.2109 | 2.038 | 0.1536 | 1.7793 |
| Interior: $\left(W_{area,\,avg}/W_{line,\,avg}\right)$ | 0.741 | 0.742 | 0.740 | 0.737 | 0.740 | 0.724 |
| Interior: $\left(W_{area,\,avg}/W_{line,\,avg}\right)_{avg}$ | 0.741 | | 0.739 | | 0.732 | |
| Annular: $W_{area,\,avg}$ | 0.1256 | 1.3400 | 0.1961 | 1.7496 | 0.2467 | 1.8654 |
| Annular: $W_{line,\,avg}$ | 0.1310 | 1.3923 | 0.2211 | 1.9041 | 0.2810 | 2.0530 |
| Annular: $\left(W_{area,\,avg}/W_{line,\,avg}\right)$ | 0.959 | 0.962 | 0.887 | 0.919 | 0.878 | 0.909 |
| Annular: $\left(W_{area,\,avg}/W_{line,\,avg}\right)_{avg}$ | 0.961 | | 0.903 | | 0.893 | |

Two additional CFD simulations were conducted to explore the appropriateness of the laminar model in interpreting the LDA results. These simulations are summarized in Table 4.4. Both of these turbulent simulations give similar area to line ratios as those determined from the laminar model. Since the Reynolds numbers for the rod bundle in the current study are within the accepted laminar regime as defined in Cheng and Todreas, the laminar model results are used in subsequent sections for interpretation of the LDA data.[6]

**Table 4.4    Velocity values from transitional and turbulent CFD simulations of the 217.5 mm storage cell.**

| | 217.5 mm Cell | |
|---|---|---|
| Turbulence model | k-$\omega$ | Realizable k-$\varepsilon$ |
| $Re$ | 900 | 900 |
| $W_{inlet}$ (m/s) | 1.6 | 1.6 |
| Interior: $W_{area,\,avg}$ | 1.7667 | 1.7400 |
| Interior: $W_{line,\,avg}$ | 2.1963 | 2.1872 |
| Interior: $\left(W_{area,\,avg}/W_{line,\,avg}\right)$ | 0.804 | 0.795 |
| Annular: $W_{area,\,avg}$ | 1.3445 | 1.3553 |
| Annular: $W_{line,\,avg}$ | 1.4000 | 1.3800 |
| Annular: $\left(W_{area,\,avg}/W_{line,\,avg}\right)$ | 0.960 | 0.982 |

The volumetric flow rate of the bundle is estimated by taking the product of the bundle area, the integrated line average of the LDA data in the bundle, and the ratio of the area to line average velocity from the CFD calculations (see $\left(W_{area,\,avg}/W_{line,\,avg}\right)_{avg}$ in Table 4.3) as shown in Equation 10.

$$Q_{bundle} = A_{bundle} \cdot W_{bundle, LDA} \cdot \left( \frac{W_{area, avg}}{W_{line, avg}} \right)_{avg, CFD\text{-}int} \qquad\qquad \textbf{10}$$

Similarly, the volumetric flow in the annulus is estimated by taking the product of the annular area, the integrated line average LDA velocity in the annulus, and the CFD-derived area to line ratio for the annular region as shown in Equation 11.

$$Q_{ann} = A_{ann} \cdot W_{ann, LDA} \cdot \left( \frac{W_{area, avg}}{W_{line, avg}} \right)_{avg, CFD\text{-}ann} \qquad\qquad \textbf{11}$$

Finally, an equivalent, assembly-averaged velocity (*i.e.* inlet velocity assuming uniform distribution) is derived from these two calculated quantities (Equation 12) to compare with the value measured directly from the mass flow controllers (Equation 7). The value calculated in Equation 7 is termed the measured, assembly average velocity. Comparison of these values gives an estimate of the error in the flow partitioning calculated in these analyses.

$$W_{avg, equiv} = \frac{Q_{bundle} + Q_{ann}}{A_{tot}} \qquad\qquad \textbf{12}$$

## 4.1 226.6 mm Storage Cell

### 4.1.1 Mid-Bundle Velocity Measurements

The velocity measurements detailed in this section were taken at $y = 94.5$ mm and $z = 1.537$ m in between spacers "H" and "I" (see Figure 4.2) for the largest storage cell (226.6 mm). This location corresponds to the middle of a long bundle run in between the two rod banks closest to the storage cell wall. Figure 4.3 shows a typical, normalized velocity profile taken inside the assembly. This profile was acquired for $Re = 50$ and was normalized by the assembly average velocity. The definitions of the bundle and annular regions are marked on the graph ($x = 100.8$ mm). The dashed black lines indicate the LDA integrated line average velocities for each region. A no-slip condition was added to the profile at the storage cell wall for all velocity profiles in this report. The velocities display a periodicity that corresponds to the rod pitch of the assembly, the maxima occurring in the centers of interstitial spaces and the minima at the narrowest points between the rods. The integrated average velocity in the annulus is over twice that of the assembly average, indicating a substantial amount of flow in the annulus despite accounting for only 28.2% of the flow area.

Figure 4.4 and Figure 4.5 show the normalized velocity distribution in the assembly for $Re = 100$ and $Re = 200$, respectively. These profiles are quite similar to that presented in Figure 4.3. Again, the annular average velocity is nearly twice that of the assembly average. The velocity profile is also periodic with the rod pitch in the assembly. These two profiles imply that the flow in the annulus is increasing slightly with increasing Reynolds number.

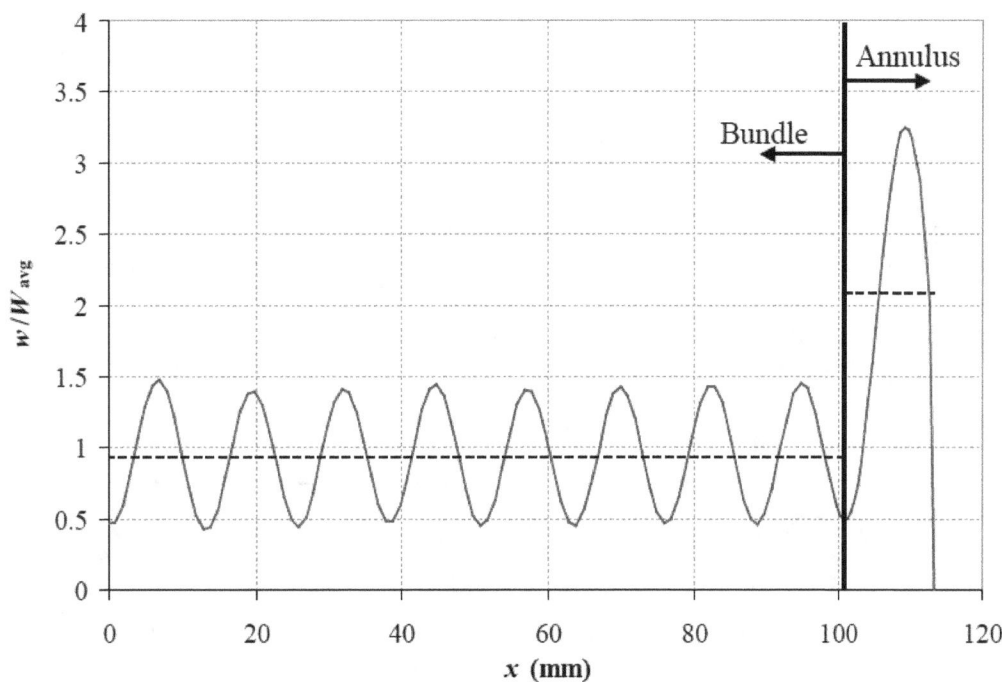

Figure 4.3    Normalized velocity as a function of position inside the assembly in the 226.6 mm storage cell for *Re* = 50.  Note: the definitions of the bundle and annular regions are shown on the graph.

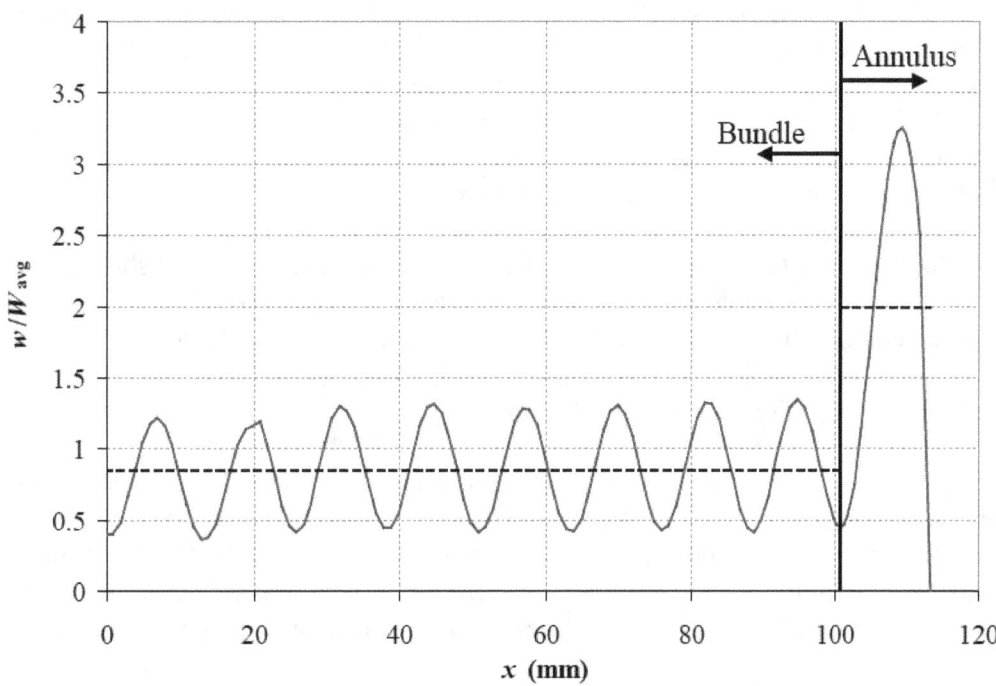

Figure 4.4    Normalized velocity as a function of position inside the assembly in the 226.6 mm storage cell for *Re* = 100.

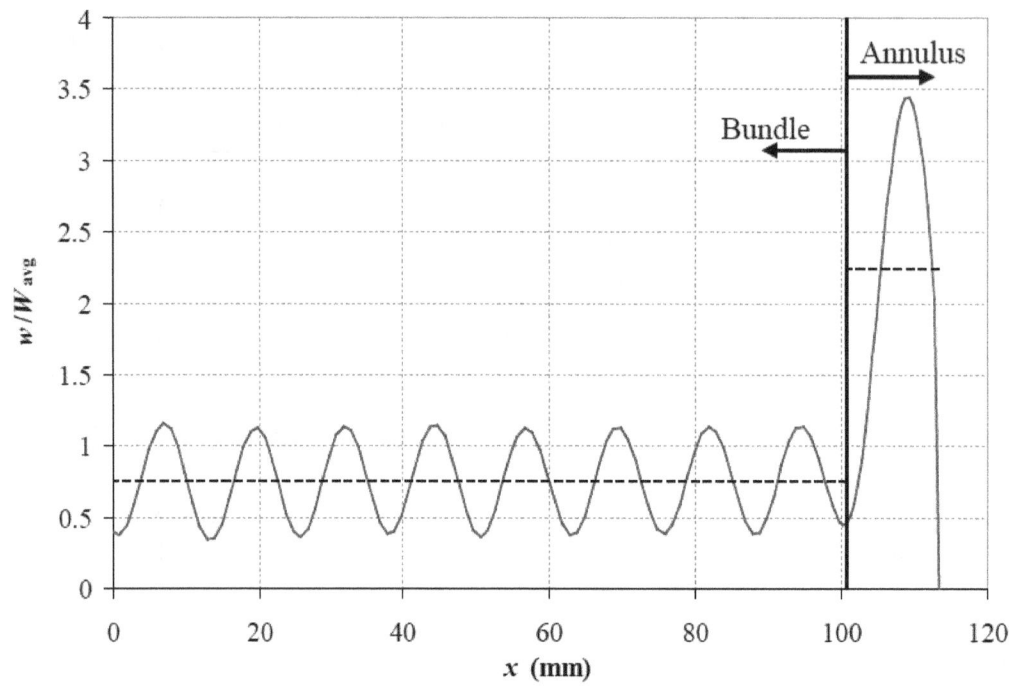

**Figure 4.5** **Normalized velocity as a function of position inside the assembly in the 226.6 mm storage cell for** *Re* **= 200.**

As Reynolds number continues to increase above *Re* = 200, a larger portion of the flow begins to partition towards the bundle. Figure 4.6 gives the normalized velocity profile for *Re* = 350. Possible shear effects are evident near the transition from the bundle to the annulus. The minimum velocity that existed at this location is now greater than the assembly average velocity and is also shifted further into the bundle. The integrated average velocities in the bundle and annulus indicate the flow partition is biasing towards the bundle.

Figure 4.7 illustrates the same trends in the velocity profile for *Re* = 500 as shown in Figure 4.6. The first minimum closest to the cell wall is again greater than the assembly average velocity and slightly displaced into the bundle, away from the location of smallest rod separation. The integrated average velocity in the bundle is greater than the assembly average velocity, and the normalized annular velocity continues to decrease.

Figure 4.8 gives the normalized velocity profile for the highest Reynolds number observed in the LDA measurements, *Re* = 900. This profile displays similar behavior to those for Reynolds numbers of 350 and above. The minimum at the boundary of the bundle and annulus is again greater than the assembly average velocity. As the Reynolds number increases, the normalized average annular and bundle velocities are decreasing and increasing, respectively.

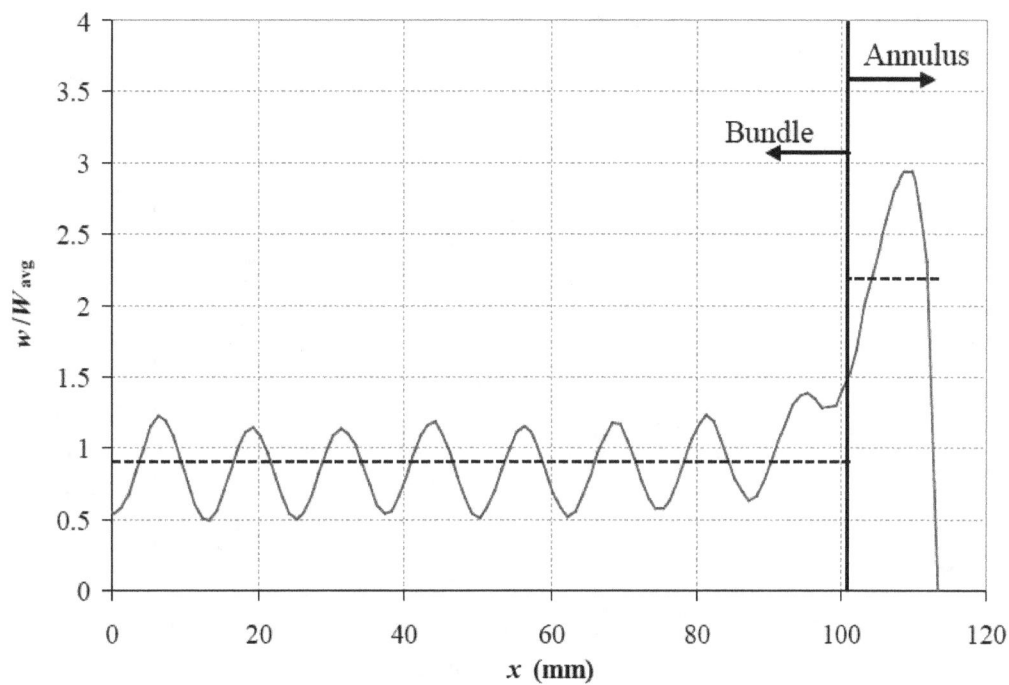

**Figure 4.6** Normalized velocity as a function of position inside the assembly in the 226.6 mm storage cell for *Re* = 350.

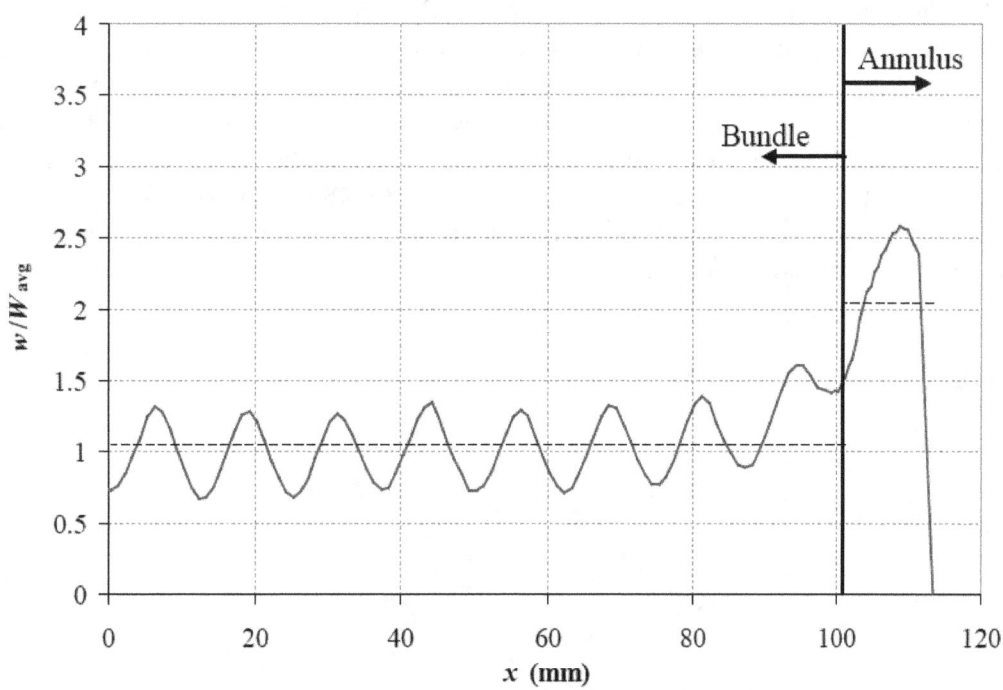

**Figure 4.7** Normalized velocity as a function of position inside the assembly in the 226.6 mm storage cell for *Re* = 500.

41

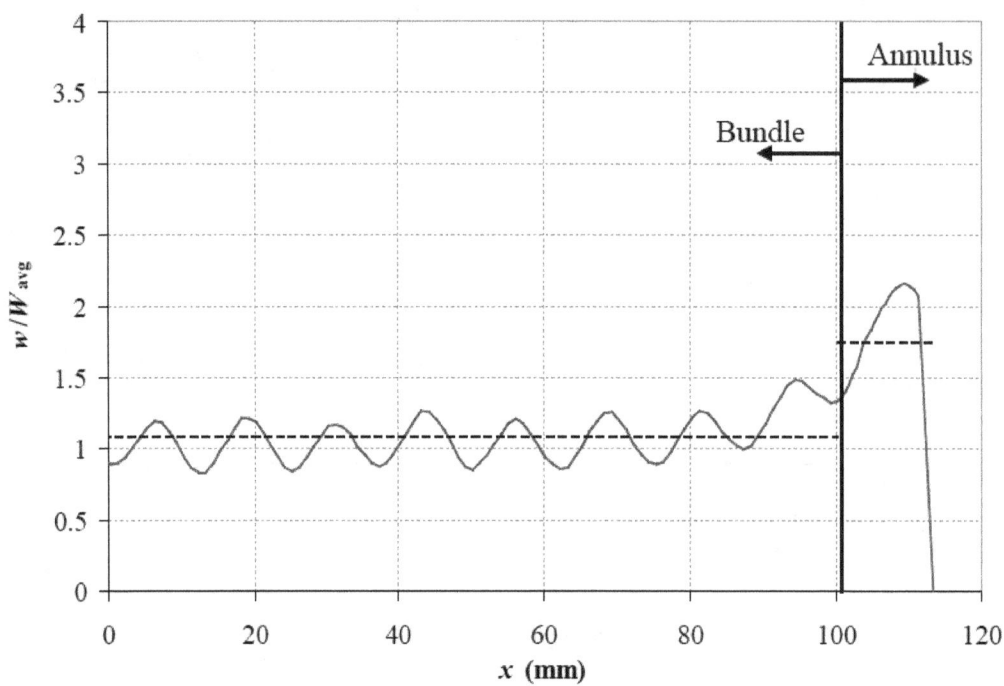

**Figure 4.8**   **Normalized velocity as a function of position inside the assembly in the 226.6 mm storage cell for** *Re* = 900.

Figure 4.9 shows the percentages of the total flow in the bundle (blue diamonds) and the annulus (red squares) as a function of Reynolds number. These values were determined using Equation 10 and Equation 11. The amount of flow initially increases in the annulus up to *Re* = 200 and then begins to decrease. The maximum flow in the annulus is 58.7% with 41.3% in the bundle. The maximum flow in the bundle is 56.3% with 43.7% in the annulus. The average flow percentages over the entire Reynolds number range are 48.6 and 51.4% in the bundle and annulus, respectively. These results indicate that the flow is nearly divided in equal parts between the bundle and annulus for the largest storage cell (226.6 mm).

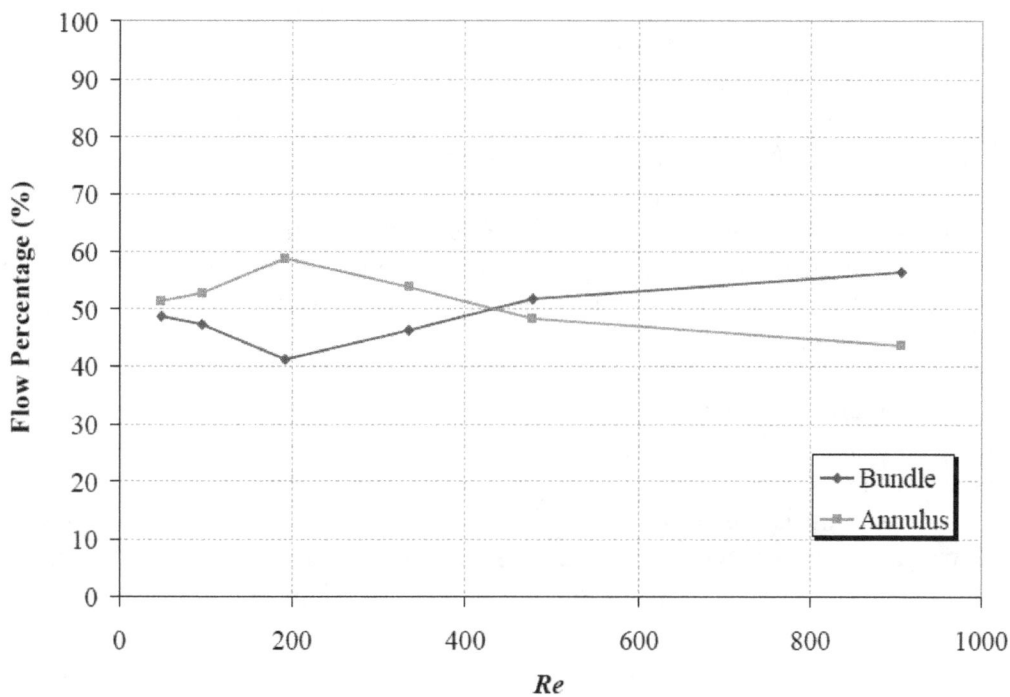

**Figure 4.9    Percentages of total flow in the bundle (blue diamonds) and in the annulus (red squares) as a function of Reynolds number for the 226.6 mm storage cell.**

Table 4.5 gives the comparison between the measured and equivalent, assembly average velocities. The absolute maximum and minimum errors are 5.4 and 0.6%, respectively. These velocities are all within the experimental uncertainty of 0.012 m/s except for $Re$ = 350 and 500. Although the partitions of flow should be viewed with a certain level of uncertainty because they are subject to the validity of a number of assumptions, the reasonable comparison of the measured and equivalent assembly velocities does suggest that the divisions of flow reported are confirmed to some extent.

**Table 4.5    Comparison of measured and equivalent, assembly average velocities for the 226.6 mm storage cell.**

| $Re$ | $W_{avg, meas}$ (m/s) | $W_{avg, equiv.}$ (m/s) | Error (%) |
|---|---|---|---|
| 50 | 0.073 | 0.074 | 1.6 |
| 100 | 0.146 | 0.138 | -5.4 |
| 200 | 0.292 | 0.281 | -3.7 |
| 350 | 0.511 | 0.525 | 2.7 |
| 500 | 0.730 | 0.778 | 6.6 |
| 900 | 1.388 | 1.395 | 0.6 |

## 4.1.2    Pre-Spacer Velocity Measurements

The velocity measurements detailed in this section were taken at $y$ = 94.5 mm and $z$ = 2.787 m upstream of spacer "D" (Figure 4.2) for the largest storage cell (226.6 mm). This location corresponds to the end of a bundle run in between the two rod banks closest to the storage cell wall. Figure 4.10 shows the normalized velocity profile for $Re$ = 100. As in the previous section, the local velocity was normalized by the assembly average velocity. The definitions of the bundle and annular regions are marked on the graph ($x = 100.8$ mm). The dashed black lines indicate the LDA integrated line average velocities for each region. The velocities display a periodicity that corresponds to the rod pitch of the assembly, the maxima occurring in the centers of interstitial spaces and the minima at the narrowest points between the rods. The integrated average velocity in the annulus is higher prior to the spacer than at the mid-bundle location shown in Figure 4.4, suggesting that the flow is redistributing preferentially to the annulus. This result is not surprising because the spacer presents a flow contraction and would naturally deflect the approaching streamlines away from the bundle into a less constrained flow path, especially at the lowest Reynolds number.

Figure 4.11 and Figure 4.12 show the normalized velocity distribution in the assembly for $Re = 350$ and $Re = 900$, respectively. As in the mid-bundle profiles, these velocity measurements show evidence of the annular flow influencing the bundle flow near the partition. The minimum bundle flow at this point is greater than the assembly average velocity. The sample rate for the LDA was lower than nominal when capturing the annular region of the velocity profile in Figure 4.11, which is the reason for the noise in the measurements. This condition was corrected once inside the bundle.

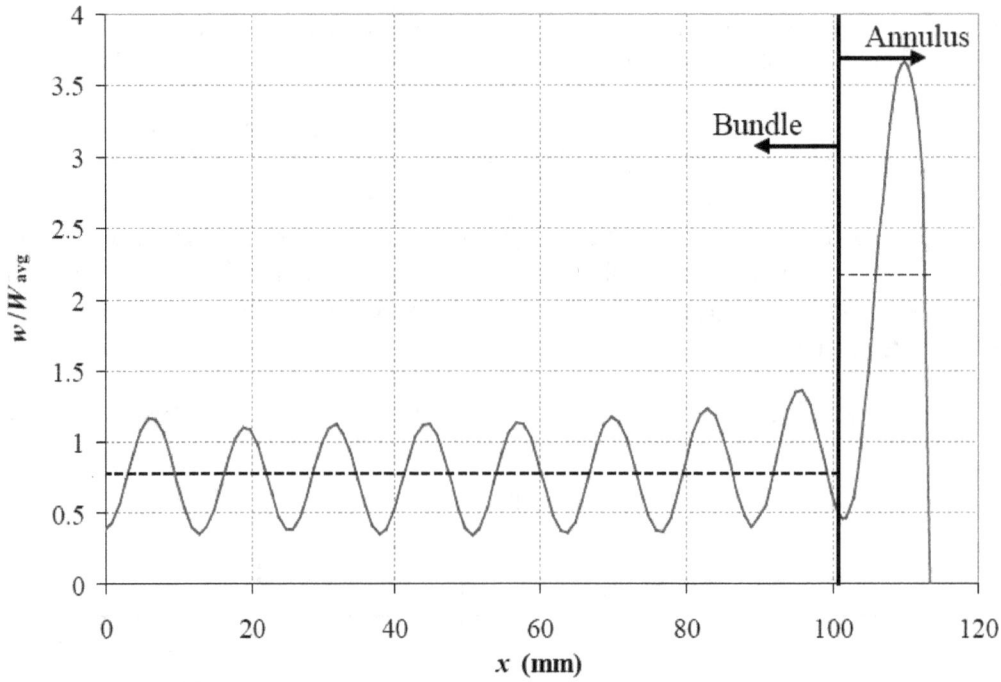

**Figure 4.10  Normalized velocity as a function of position inside the assembly in the 226.6 mm storage cell before a spacer for $Re = 100$.**

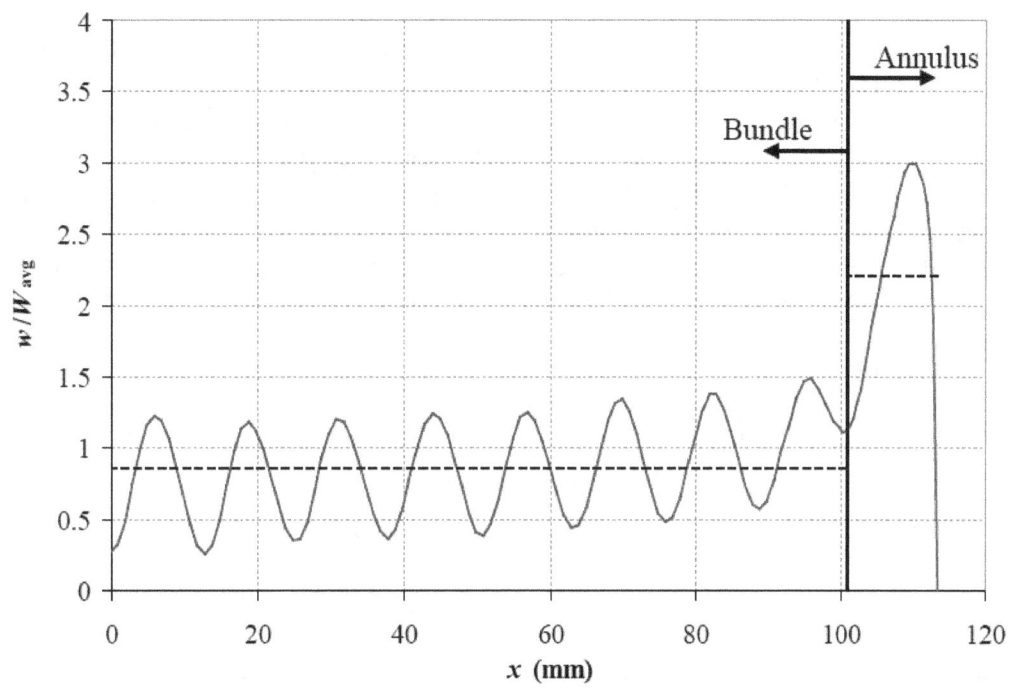

Figure 4.11  Normalized velocity as a function of position inside the assembly in the 226.6 mm storage cell before a spacer for *Re* = 350.

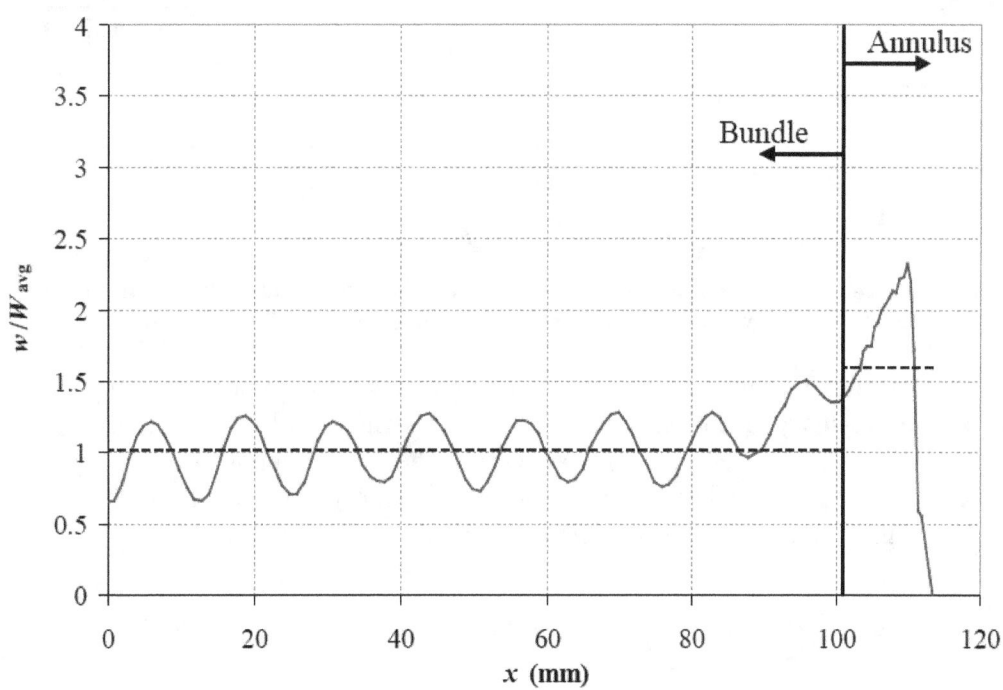

Figure 4.12  Normalized velocity as a function of position inside the assembly in the 226.6 mm storage cell before a spacer for *Re* = 900.

The percentages of flow in the annulus and bundle at a $z$-location corresponding to just upstream of a spacer are presented in Figure 4.13. A redistribution of flow into the annulus is evident at the lowest Reynolds number, indicating the spacer may affect the upstream flow distribution. However, these effects are not as marked for $Re > 200$. Also, the CFD results used to interpret the LDA data were obtained for a fully developed laminar condition. This assumption of fully developed flow may not be appropriate with the addition of a flow contraction upstream of the measurements. The average flow percentages over the entire Reynolds number range are 48.4 and 51.6% in the bundle and annulus, respectively. These results again indicate that the flow is nearly divided in equal parts between the bundle and annulus for the largest storage cell (226.6 mm).

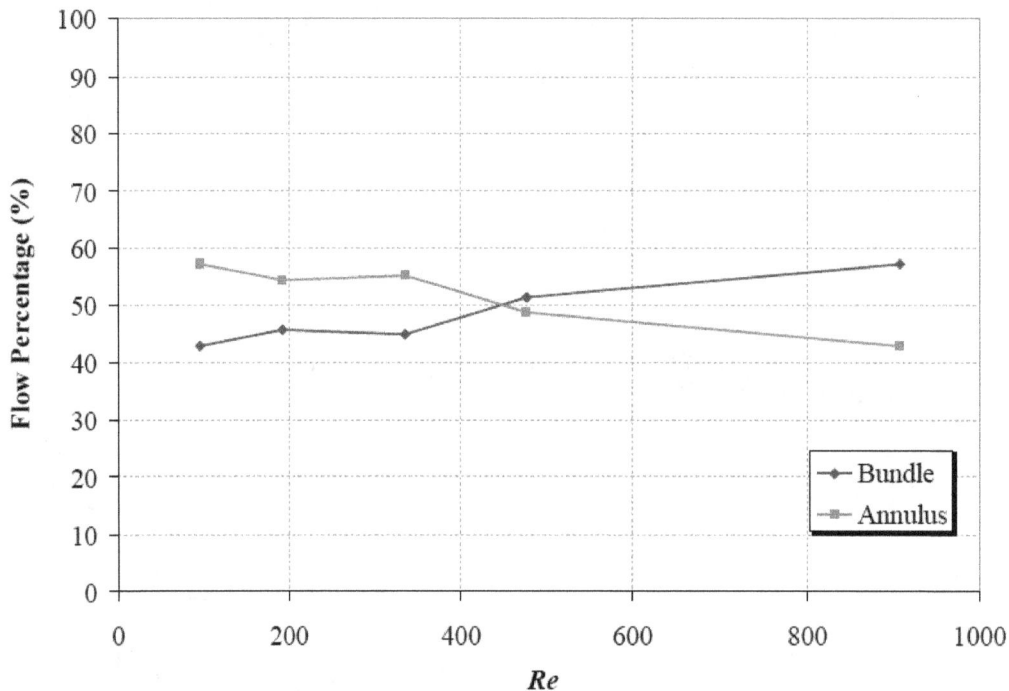

**Figure 4.13**     **Percentages of total flow in the bundle (blue diamonds) and in the annulus (red squares) as a function of Reynolds number for the 226.6 mm storage cell before a spacer.**

Table 4.6 gives the comparison between the measured and equivalent, assembly average velocities at the pre-spacer location. The absolute maximum and minimum errors are 25.2 and 0.5%, respectively. These velocities are all within the experimental uncertainty of 0.012 m/s except for $Re = 200$ and 900. These comparisons again show reasonable agreement, particularly when considering the presence of the spacer may affect the distribution of flow at this location.

**Table 4.6**   Comparison of measured and equivalent assembly average velocities for the 226.6 mm storage cell before a spacer.

| *Re* | $W_{avg, meas}$ (m/s) | $W_{avg, equiv.}$ (m/s) | Error (%) |
|------|-----------------------|--------------------------|-----------|
| 100  | 0.146                 | 0.140                    | -4.3      |
| 200  | 0.292                 | 0.219                    | -25.2     |
| 350  | 0.511                 | 0.514                    | 0.5       |
| 500  | 0.730                 | 0.721                    | -1.2      |
| 900  | 1.388                 | 1.300                    | -6.3      |

### 4.1.3   Post-Spacer Velocity Measurements

The velocity measurements detailed in this section were taken at $y = 94.5$ mm and $z = 1.326$ m downstream of spacer "I" (see Figure 4.2) for the largest storage cell (226.6 mm). This location corresponds to the beginning of a bundle run in between the two rod banks closest to the storage cell wall. Figure 4.14 shows the normalized velocity profile for $Re = 100$. As in previous sections, the assembly average velocity was used to normalize the local velocity. The definitions of the bundle and annular regions are marked on the graph ($x = 100.8$ mm). The dashed black lines indicate the LDA integrated line average velocities for each region. The velocities display a periodicity that corresponds to the rod pitch of the assembly, the maxima occurring in the centers of interstitial spaces and the minima at the narrowest points between the rods. The annulus to bundle ratio of integrated average velocity is considerably higher after the spacer than at the mid-bundle location, again suggesting that the flow is redistributing preferentially to the annulus at the spacers. The spacers present a flow contraction and would naturally deflect the approaching streamlines away from the bundle. The spacers are also designed to promote flow mixing in order to enhance heat transfer, which will strongly affect the flow distribution.

Figure 4.15 and Figure 4.16 show the normalized velocity distribution in the assembly for $Re = 350$ and $Re = 900$, respectively. These velocity measurements show evidence of the wake effects from the spacer. These effects are observed for $Re \geq 350$ at this $z$-location. The initial effects are broadening and enhancement of every other maximum in the bundle velocity profile as shown in Figure 4.15. As Reynolds number increases, these wake structures become more pronounced as seen in Figure 4.16. This velocity profile may appear chaotic at first glance but is actually periodic at a frequency of twice the pitch, which corresponds to the repeating pattern of small turning vanes within the spacer.

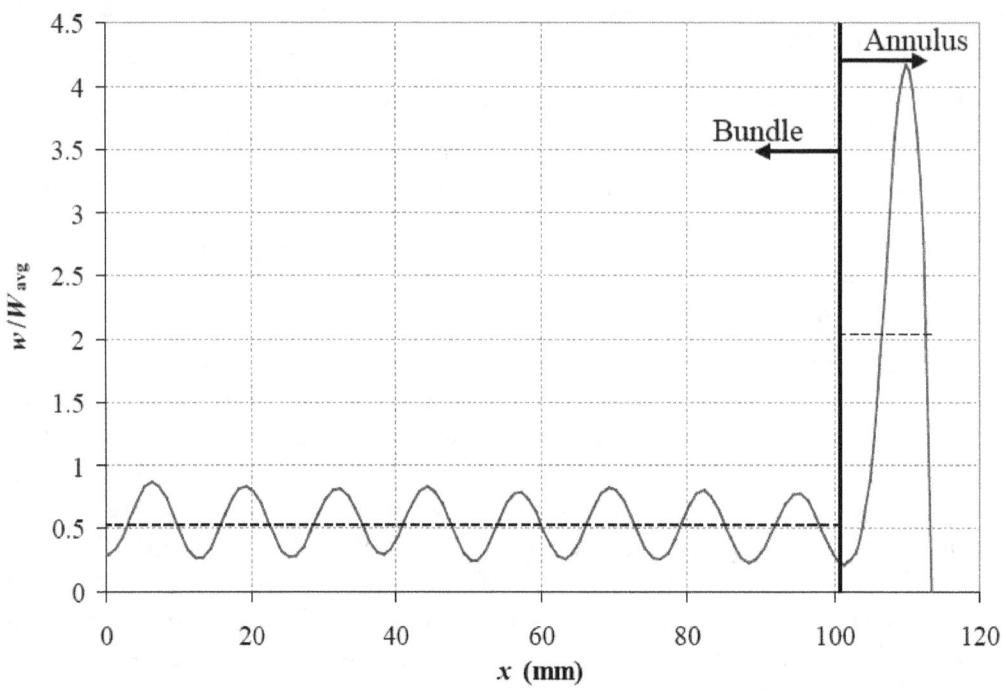

**Figure 4.14** Normalized velocity as a function of position inside the assembly in the 226.6 mm storage cell after a spacer for $Re = 100$.

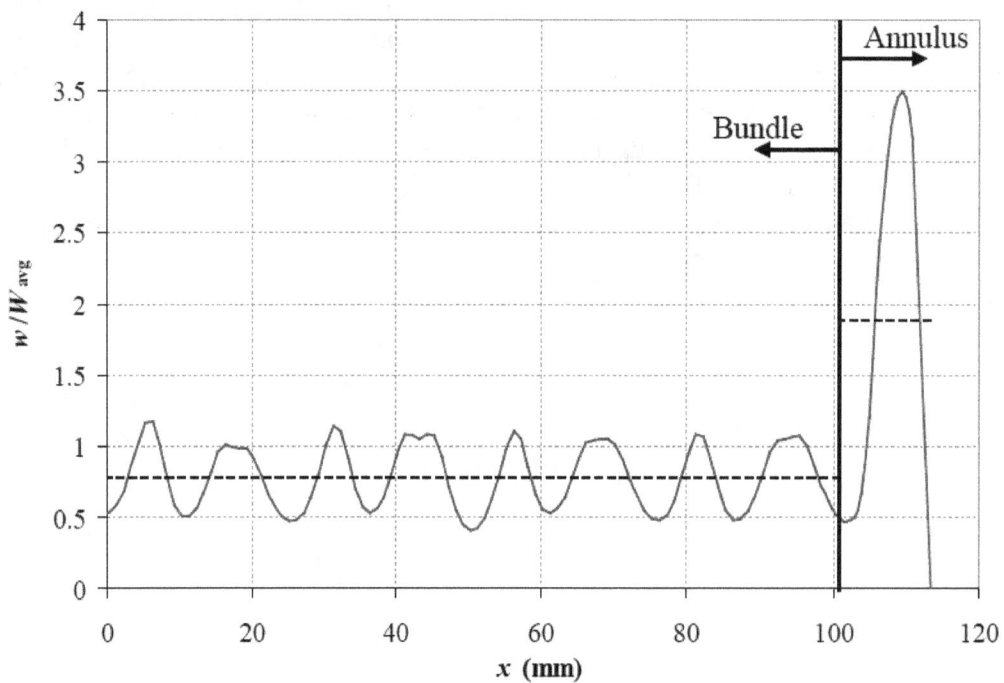

**Figure 4.15** Normalized velocity as a function of position inside the assembly in the 226.6 mm storage cell after a spacer for $Re = 350$.

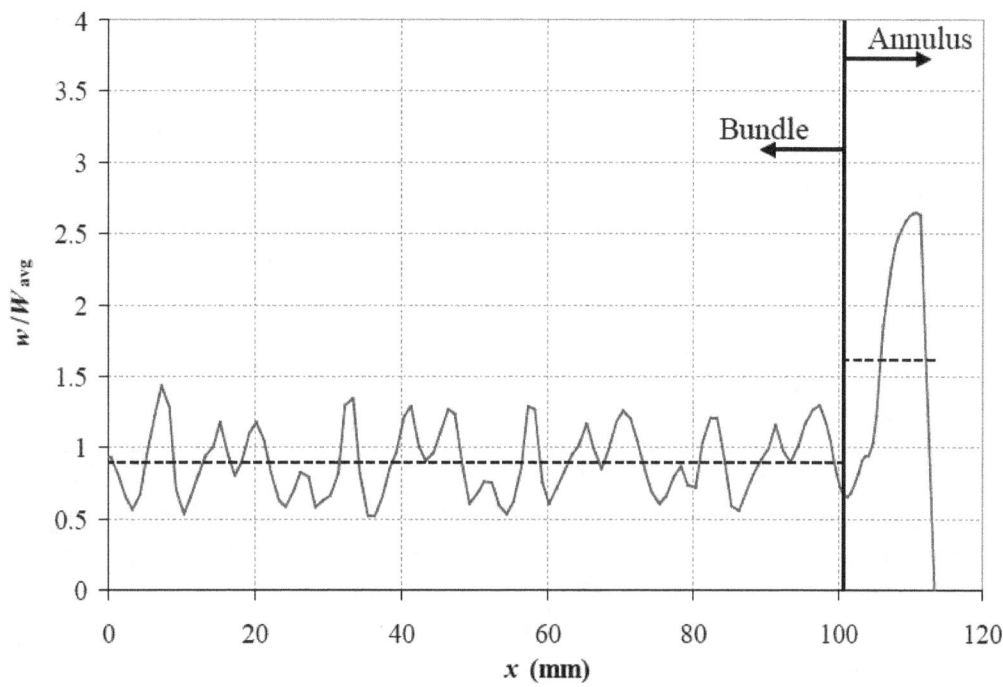

**Figure 4.16  Normalized velocity as a function of position inside the assembly in the 226.6 mm storage cell after a spacer for *Re* = 900.**

Figure 4.17 shows the flow percentages in the bundle and annulus as calculated at the post-spacer location for the largest storage cell (226.6 mm). These results suggest that the flow does redistribute into annulus around the spacers, particularly at lower Reynolds numbers. The average flow percentages over the entire Reynolds number range are 43.1 and 56.9% in the bundle and annulus, respectively. This flow partition indicates slightly higher flow in the annulus compared to mid-bundle location, 51.4%. However, the assumption of fully developed laminar flow in the CFD results is likely inappropriate as evidenced in the higher Reynolds velocity profiles and will lead to errors in the interpretation of the LDA data.

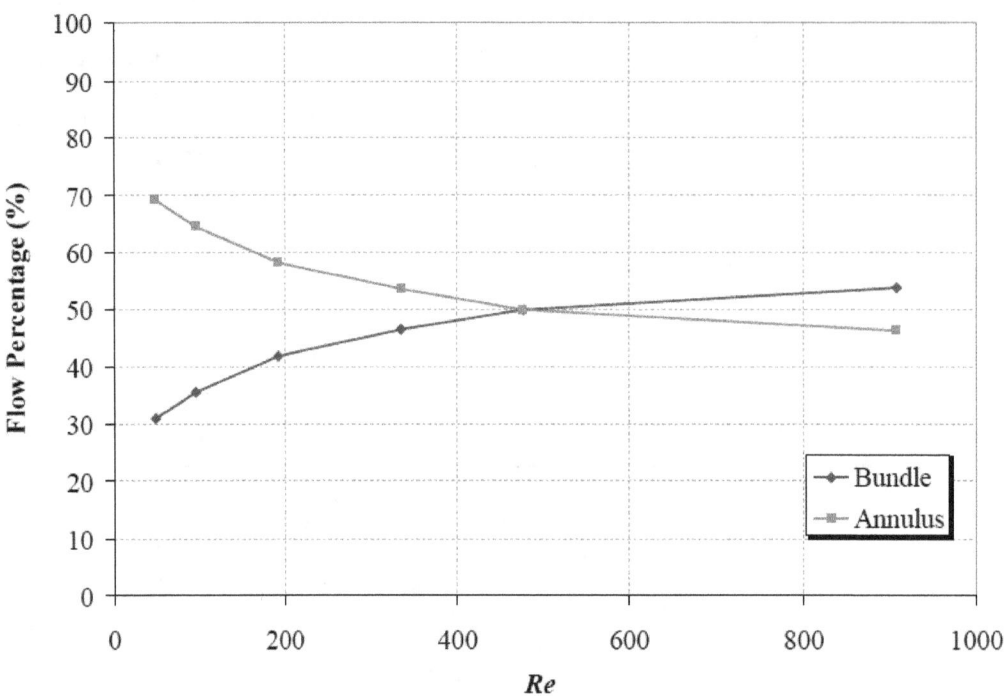

**Figure 4.17  Percentages of total flow in the bundle (blue diamonds) and in the annulus (red squares) as a function of Reynolds number for the 226.6 mm storage cell after a spacer.**

Table 4.7 gives the comparison between the measured and equivalent, assembly average velocities at the post-spacer location for the largest storage cell (226.6 mm). The absolute maximum and minimum errors are 20.2 and 10%, respectively. All the equivalent velocities are lower than the measured values. This discrepancy is most likely due to the use of fully developed CFD results to interpret velocity profiles that exhibit obvious wake effects from the spacer.

**Table 4.7  Comparison of measured and equivalent assembly average velocities for the 226.6 mm storage cell after a spacer.**

| $Re$ | $W_{avg, meas}$ (m/s) | $W_{avg, equiv.}$ (m/s) | Error (%) |
|---|---|---|---|
| 50 | 0.073 | 0.059 | -18.7 |
| 100 | 0.146 | 0.117 | -20.2 |
| 200 | 0.292 | 0.234 | -19.9 |
| 350 | 0.511 | 0.454 | -11.2 |
| 500 | 0.730 | 0.657 | -10 |
| 900 | 1.388 | 1.221 | -12 |

## 4.2   221.8 mm Storage Cell

### 4.2.1  Mid-bundle Velocity Measurements

The velocity measurements detailed in this section were taken at $y$ = 94.5 mm and $z$ = 1.537 m in between spacers "H" and "I" (see Figure 4.2) for the middle-sized storage cell (221.8 mm). This location corresponds to the middle of a long bundle run in between the two rod banks closest to the storage cell wall. Figure 4.18 shows the normalized velocity profile taken inside the assembly for $Re = 100$. The definitions of the bundle and annular regions are marked on the graph ($x = 100.8$ mm). The dashed black lines indicate the LDA integrated line average velocities for each region. The ratio of the bundle to annulus integrated average velocity is greater than in the 226.6 mm storage cell, indicating more flow is now passing through the bundle rather than the annulus.

Figure 4.19 and Figure 4.20 show the normalized velocity distribution in the assembly for $Re = 350$ and $Re = 900$, respectively. These profiles also indicate that more flow is passing through the bundle than the annulus as compared to the 226.6 mm storage cell. The shear effects observed in Figure 4.6 at the bundle/annulus transition are not evident in Figure 4.19. These effects are apparent at the higher Reynolds number flow as shown in Figure 4.20, although not as pronounced.

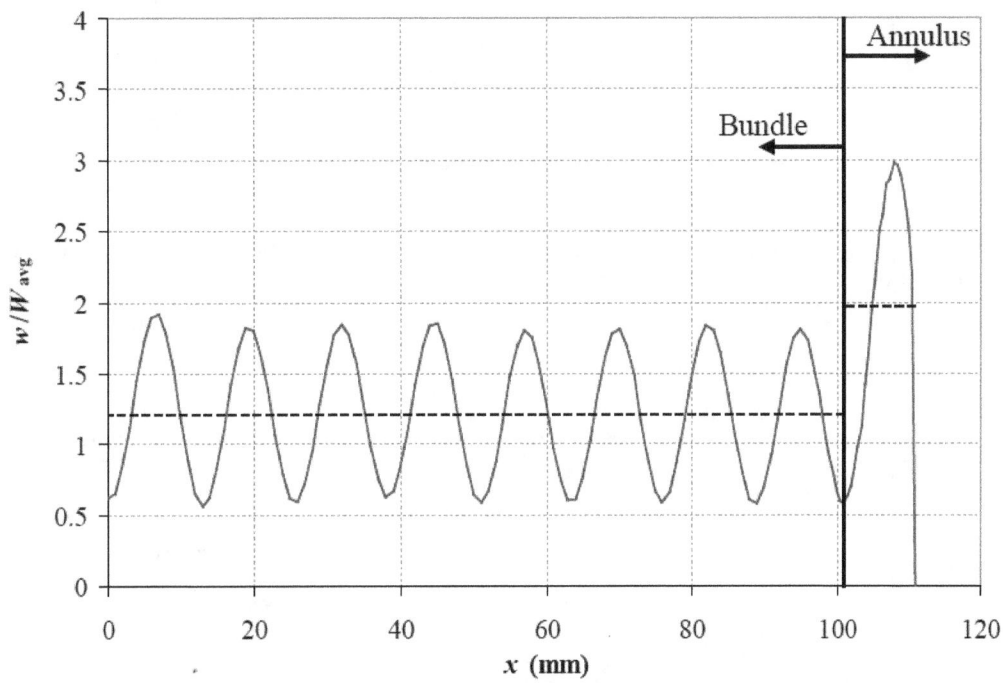

**Figure 4.18  Normalized velocity as a function of position inside the assembly in the 221.8 mm storage cell for**
**$Re = 100$.**

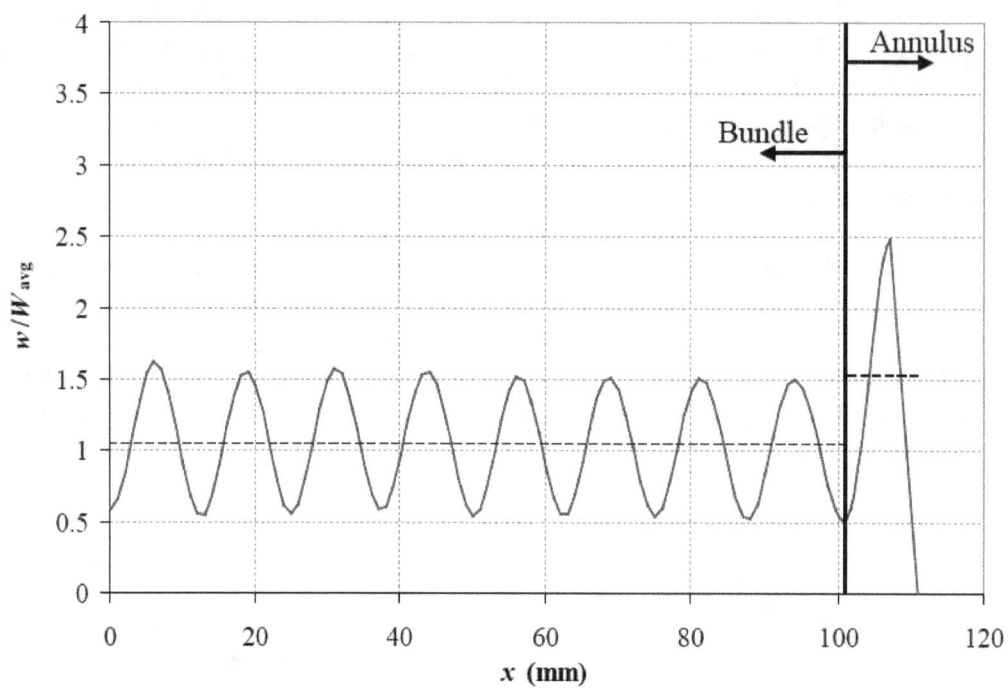

**Figure 4.19** Normalized velocity as a function of position inside the assembly in the 221.8 mm storage cell for
$Re = 350$.

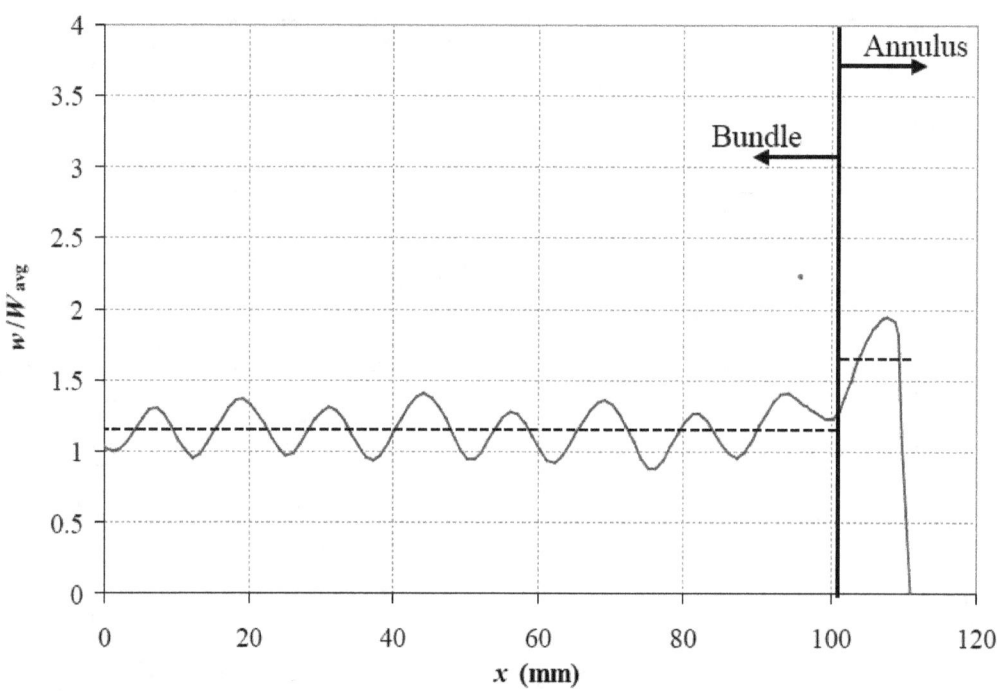

**Figure 4.20** Normalized velocity as a function of position inside the assembly in the 221.8 mm storage cell for
$Re = 900$.

Figure 4.21 shows the percentages of the total flow in the bundle (blue diamonds) and the annulus (red squares) as a function of Reynolds number for the 221.8 mm storage cell. These values were determined using Equation 10 and Equation 11. The amount of flow is relatively constant across the entire flow range. The average flow percentages over the entire Reynolds number range are 65.0 and 35.0% in the bundle and annulus, respectively. These results indicate that more flow passes through the bundle than the annulus for the middle-sized storage cell (221.8 mm).

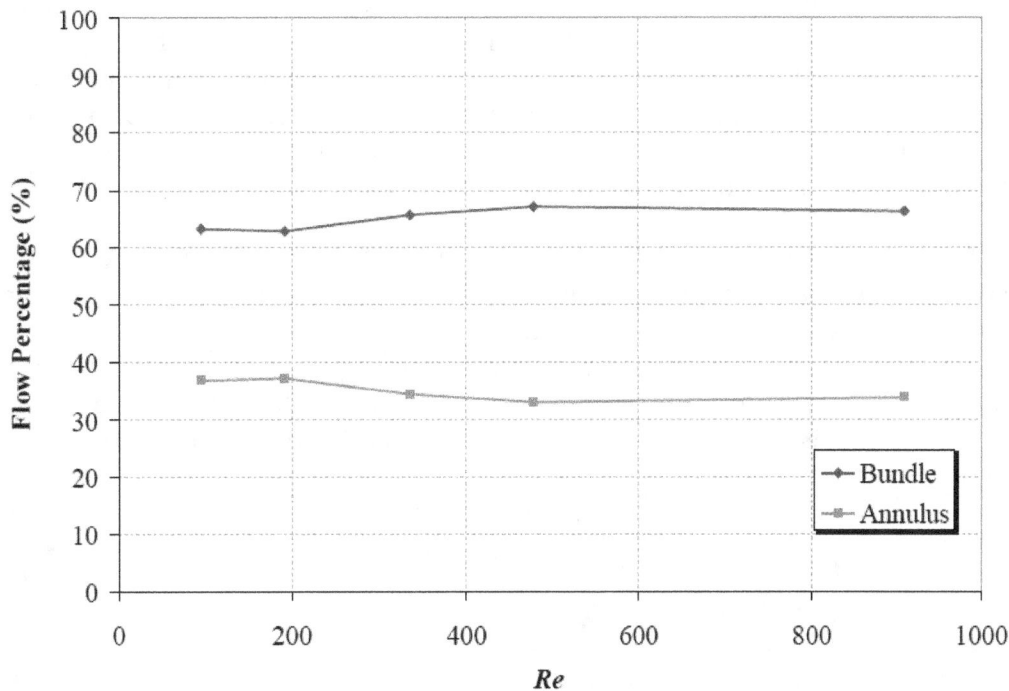

**Figure 4.21  Percentages of total flow in the bundle (blue diamonds) and in the annulus (red squares) as a function of Reynolds number for the 221.8 mm storage cell.**

Table 4.8 gives the errors between the measured and equivalent average velocities for the 221.8 mm storage cell. The average error of the equivalent velocity is -5.5%. While all the equivalent velocities are outside the experimental uncertainty with the exception of $Re = 900$, the relative agreement is again viewed as confirmation of the flow percentages calculated previously.

**Table 4.8   Comparison of measured and equivalent assembly average velocities for the 221.8 mm storage cell.**

| $Re$ | $W_{avg, meas}$ (m/s) | $W_{avg, equiv.}$ (m/s) | Error (%) |
|------|------------------------|--------------------------|-----------|
| 50   | 0.079                  | 0.064                    | -19.7     |
| 100  | 0.158                  | 0.174                    | 9.7       |
| 200  | 0.317                  | 0.302                    | -4.5      |
| 350  | 0.554                  | 0.506                    | -8.8      |
| 500  | 0.792                  | 0.715                    | -9.6      |
| 900  | 1.504                  | 1.504                    | 0.0       |

## 4.3    217.5 mm Storage Cell

### 4.3.1    Mid-bundle Velocity Measurements

The velocity measurements detailed in this section for Reynolds numbers of 50, 100, 350, and 900 were taken at $y = 94.5$ mm and $z = 1.537$ m in between spacers "H" and "I" (see Figure 4.2) for the 217.5 mm storage cell. The measurements for $Re = 200$ and 500 were taken at $y = 18.9$ mm and the same $z$-location. Due to the close proximity of the annular wall to the first rod bank and the resulting reflections, velocity measurements were not obtained in the annulus for the smallest storage cell. However, the annular velocity is estimated by subtracting the bundle flow from the assembly total and dividing by the annulus area. Figure 4.22 shows the normalized velocity profile in the 217.5 mm storage cell for $Re = 100$. Logically, the integrated average velocity in the bundle is greater than that observed in the two larger cells. This profile displays the same periodicity shown in previous sections. However, the difference in the maximum to minimum in the bundle is greater for the smallest cell.

Figure 4.23 gives the normalized velocity profile for $Re = 350$. The average velocity in the bundle has decreased slightly from the case of $Re = 100$, indicating more flow is going into the annulus as Reynolds number increases. This trend is also observed for $Re = 900$ in Figure 4.24. The average velocities in the annulus and bundle are nearly equal for this flow. The wake effects of the spacer are also evident in this profile indicating that the flow has not reestablished fully developed flow. These effects were not seen at the mid-bundle location for the two larger storage cells.

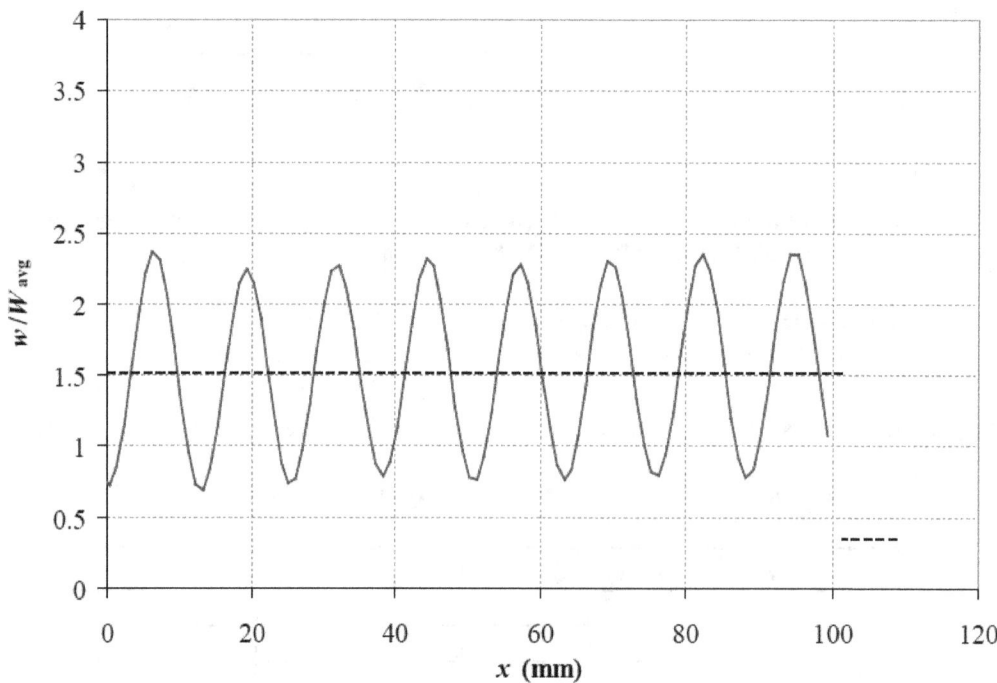

**Figure 4.22  Normalized velocity as a function of position inside the assembly in the 217.5 mm storage cell for**
        $Re = 100$.

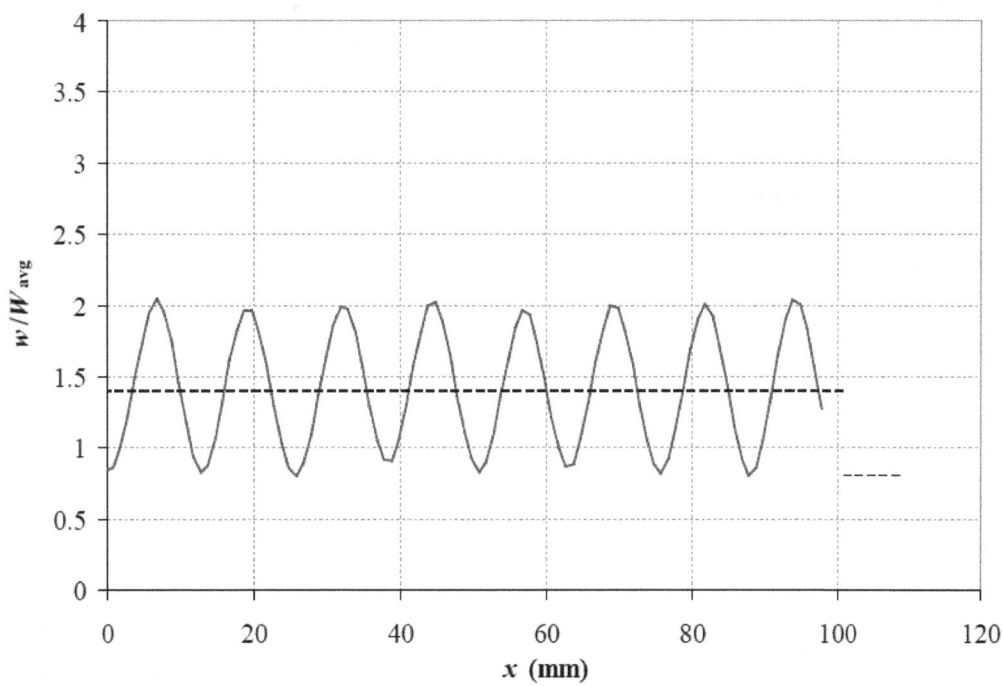

**Figure 4.23** Normalized velocity as a function of position inside the assembly in the 217.5 mm storage cell for *Re* = 350.

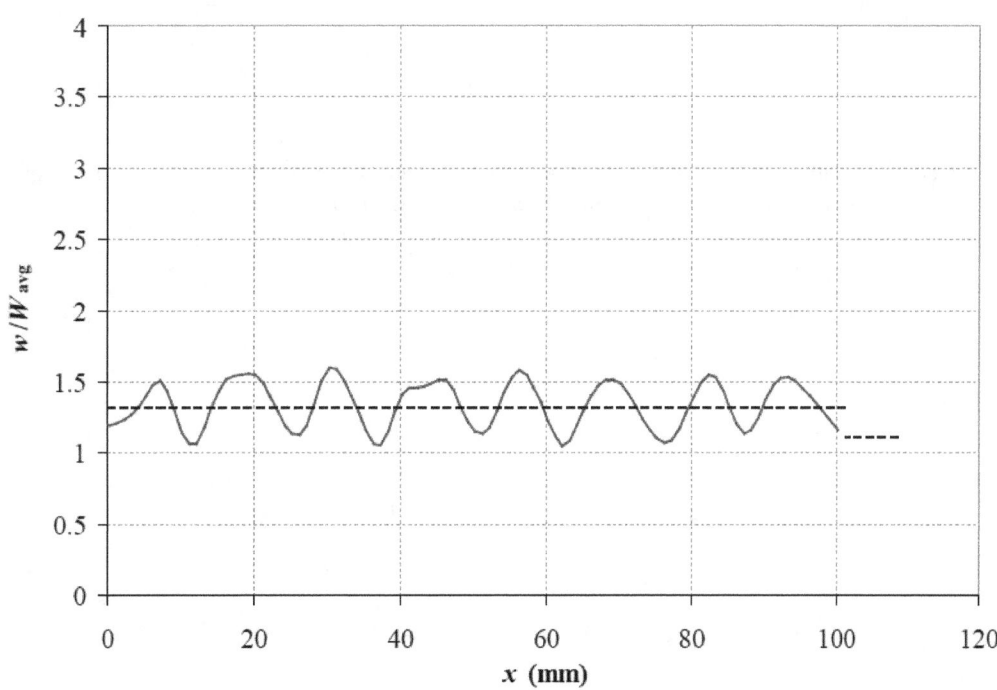

**Figure 4.24** Normalized velocity as a function of position inside the assembly in the 217.5 mm storage cell for *Re* = 900.

Figure 4.25 shows the percentages of flow in the bundle and annulus for the smallest storage cell (217.5 mm). These results indicate that flow in the bundle decreases with increasing Reynolds number. The average flow percentages were 88.8 and 11.2% in the bundle and annulus, respectively. These results are subject to greater uncertainty than the other storage cells because the annulus was not measured independently. The equivalent assembly average velocity therefore cannot be calculated.

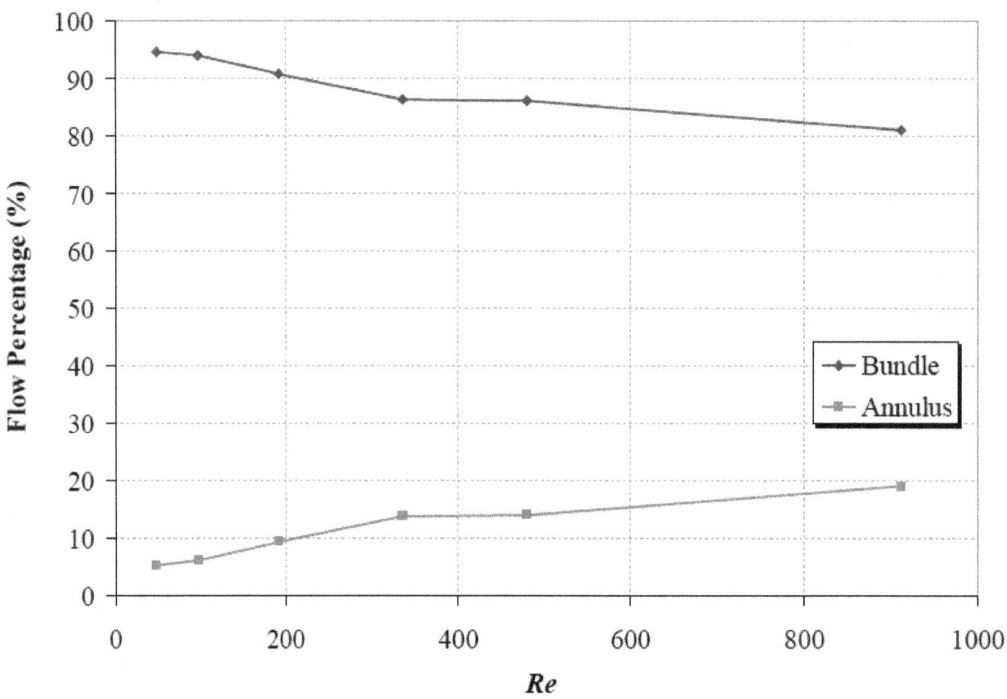

**Figure 4.25  Percentages of total flow in the bundle (blue diamonds) and in the annulus (red squares) as a function of Reynolds number in the 217.5 mm storage cell.**

# 5  SUMMARY

These studies represent the first hydraulic characterizations of a full length, highly prototypic PWR fuel assembly in low Reynolds number flows. There are at least two advantages to the testing of full scale, prototypic components. First, the use of actual hardware and dimensionally accurate geometries eliminates any issues arising from scaling arguments. Second, many of the prototypic components contain intricacies by design that would not be reproduced by using simplified flow elements. While this approach yields results that are inherently specific to the fuel assembly under testing, the differences in commercial designs are considered minor, particularly when considering the hydraulics of the entire assembly.

The generic use of best estimate flow parameters available in the literature for the hydraulic analysis of a nuclear fuel assembly may significantly underestimate the resistance to laminar flow in the assembly. This underestimation leads to an overestimate of the cooling effects of naturally induced flows that develop in dry casks under normal storage conditions or wet pool cells during complete loss of coolant accidents. Early on, the underestimation of these parameters was balanced by the large safety margin in the evaluation of peak cladding temperature due to the low decay heat. As the applicants continue to increase the stored fuel decay heat, these parameters became important to the analysis. As such, the estimation of these parameters were estimated using CFD which was validated using the data in this report.

A commercial 17×17 PWR fuel assembly was hydraulically characterized by measuring both pressure drops and velocities inside three different storage cells in the laminar regime. Two of the storage cell sizes (226.6 mm and 221.8 mm) were chosen to span the cell size commonly used in dry storage casks and the third size (217.5 mm) was chosen as a practical minimum that forced most the air flow through the tube bundle. These tests spanned Reynolds numbers from 10 to 1000 based on the hydraulic diameter and average assembly velocity. The pressure drop results were used to calculate viscous and form loss coefficients, namely $S_{LAM}$ and $\Sigma k$, respectively. The velocity profiles were used to estimate the partitioning of flow in the tube bundle and the annular region between the tube bundle and the storage cell wall. The velocity profile data also suggests that transition away from laminar flow behavior may begin at Reynolds numbers as low as 200.

## 5.1  Pressure Drop Measurements

Pressure drop measurements were collected at 52 different flow rates between 25 and 2100 slpm. Three high precision quartz crystal differential pressure gauges collected data from 36 pressure ports. The pressure ports were positioned to allow characterization of all individual spacers and bundle runs along the axis of the assembly. Overall pressure drop data was used to calculate $S_{LAM}$ and $\Sigma k$ hydraulic loss coefficients. The technique used to determine these coefficients was successfully validated by investigation of flow in a simple annulus for which an analytic value for $S_{LAM}$ is known, see Appendix A for details. Hydraulic loss coefficients were also determined for individual assembly components by the same technique and integrated over the length of the assembly to determine equivalent overall coefficients. The equivalent overall coefficients were in excellent agreement with the directly measured overall coefficients in all cases tested.

For reference, comparison was made with the BWR assembly characterized in a previous study. The BWR assembly was also re-tested, and the new pressure drop data is in excellent agreement with the previous study.

The smallest PWR storage cell tested (217.5 mm) is analogous to a BWR canister. The overall pressure drop across the PWR assembly in this storage cell was significantly greater than the overall pressure drop across the BWR assembly. The BWR pressure drop is lower because the assembly has fewer grid spacers, and partial length rods result in a significant increase in flow area in the upper third of the assembly. The overall $S_{LAM}$ and $\Sigma k$ hydraulic parameters for the PWR assembly in the 217.5 mm storage cell were determined to be 132.9 and 30.6 respectively.

The middle sized PWR storage cell (221.8 mm) tested represents the smallest cell typically used in commercial dry casks. The overall pressure drop across the PWR assembly in the 221.8 mm cell was found to be essentially the same as the overall pressure drop across the BWR assembly. The overall $S_{LAM}$ and $\Sigma k$ hydraulic parameters for the PWR assembly in the 221.8 mm storage cell were determined to be 109.9 and 27.7 respectively.

The largest PWR storage cell tested (226.6 mm) represents the largest cell typically used in commercial dry casks. The overall pressure drop across the PWR assembly in the 226.6 mm cell was significantly lower than the overall pressure drop across the BWR assembly. The overall $S_{LAM}$ and $\Sigma k$ hydraulic parameters for the PWR assembly in the 226.6 mm storage cell were determined to be 98.5 and 27.4 respectively.

The viscous loss coefficient, $S_{LAM}$, exhibits a larger dependence on storage cell hydraulic diameter than the form loss coefficient, $\Sigma k$. To aid in determining the appropriate coefficients to use with storage cell sizes not tested, empirical power law correlations were determined for $S_{LAM}$ and $\Sigma k$ as a function of storage cell hydraulic diameter. The resulting correlations based on the full range of flow rates tested ($Re = 10$ to $1000$) are:

$$S_{LAM} = 57 + 1.891E\text{-}7 \cdot D_{H,\,Ref.}^{-4.348}$$

$$\Sigma k = 9.872E\text{-}1 \cdot D_{H,\,Ref.}^{-0.7527}$$

where $D_{H,\,Ref.}$ is the storage cell hydraulic diameter in meters. The correlations should only be used for 17×17 PWR fuel assemblies with storage cells smaller than 230 mm.

## 5.2  Assembly Velocity Measurements

Velocity profiles were measured across the PWR bundle using laser Doppler anemometry (LDA) These profiles are used in estimating the flow partition between the bundle and annular regions within the assembly. These measurements also indicated a redistribution of flow after spacers and intermediate flow mixers (IFMs) at higher flow rates, suggesting significant wake effects. The wake disturbances in the flow were not apparent in the mid-bundle measurements, which may suggest that the flow has reestablished a fully developed condition.

The partitioning of flow between the bundle and annular regions showed a strong dependence on storage cell size and a weaker dependence on Reynolds number. For $Re = 400$ at the mid-bundle location, the percentage of annular flow was 12, 34, and 52% in the 217.5, 221.8, and 226.6 mm storage cells, respectively.  In general the percentage of annular flow decreased with Reynolds number except in the 217.5 mm cell where the percentage increased from 5 to 20% at $Re$ = 50 to 900, respectively.

# 6 REFERENCES

1. Lindgren, E.R. and Durbin, S.G., 2007, "Characterization of Thermal-Hydraulic and Ignition Phenomena in Prototypic, Full-Length Boiling Water Reactor Spent Fuel Pool Assemblies after a Complete Loss-of-Coolant Accident," SAND2007-2270.

2. Farell, C. and Youssef, S., 1996, "Experiments on Turbulence Management Using Screens and Honeycombs," *J. Fluids Eng.*, **118**, 26-32.

3. Durst, F., Melling, A., Whitelaw, J.H., 1981, Principles and Practice of Laser-Doppler Anemometry, New York: Academic Press.

4. Albrecht, H., Borys, M., Damaschke, N., and Tropea, C., 2003, Laser Doppler and Phase Doppler Measurement Techniques, Berlin: Spinger-Verlag.

5. Kays, W.M. and M.E. Crawford, 1980, Convective Heat and Mass Transfer, Second Edition, New York, NY: McGraw- Hill.

6. Cheng, S. and Todreas, N.E., 1986, "Hydrodynamic Models and Correlations for Bare and Wire-Wrapped Hexagonal Rod Bundles—Bundle Friction Factors, Subchannel Friction Factors and Mixing Parameters," *Nucl. Eng. Des.*, **92**, 227-251.

# APPENDIX A.  ANALYTIC VALIDATION

The technique used to determine the hydraulic loss coefficients was successfully validated by the following investigation of flow in a simple annulus for which an analytic value for $S_{LAM}$ is known.  The dimensions of the annulus were sized to represent flow in semi-infinite parallel plates and resulted in measurements that were within the normal operating range of the flow controllers and pressure gauges used in the larger PWR characterization study.  The same flow controllers, pressure gauges, and control systems were used in the both this validation exercise and the PWR (and previous BWR) characterizations.

The pressure drop in the annular flow area formed between two concentric sections of pipe was measured to compare with the known analytic solution.  The outer and inner pipes were 4 in. and 3 in. PVC, respectively.  The average inner diameter (ID) of the 4 in. pipe was measured to be 101.1 mm (3.980 in.), and the average outer diameter (OD) of the 3 in. pipe was measured to be 89.0 mm (3.502 in.).  Table A.1 summarizes the hydraulic characteristics of the annular flow region.  The annulus has a hydraulic diameter of $D_{H, Ref.} = 12.1$ mm (0.478 in.) and a flow area of 1812 mm$^2$ (2.809 in$^2$).

**Table A.1**    **Hydraulic characteristics of the annular flow region.**

| Description | Metric Value | English Value |
|---|---|---|
| ID | 89.0 mm | 3.502 in. |
| OD | 101.1 mm | 3.980 in. |
| Area | 1812 mm$^2$ | 2.809 in$^2$ |
| Wetted perimeter | 597.0 mm | 23.505 in. |
| $D_{H, Ref}$ | 12.1 mm | 0.478 in |

Figure A.1 shows the flow area that was studied.  Due to the manner of construction, the inner pipe ends 85.725 mm (3.375 in) below the outer pipe.  However, all pressure ports were located below the top of the inner pipe.  The inner pipe was centered by three screws at the top located at 120° spacing (see the left photo in Figure 1) and by a 3 in. to 4 in. reducing coupling at the bottom.

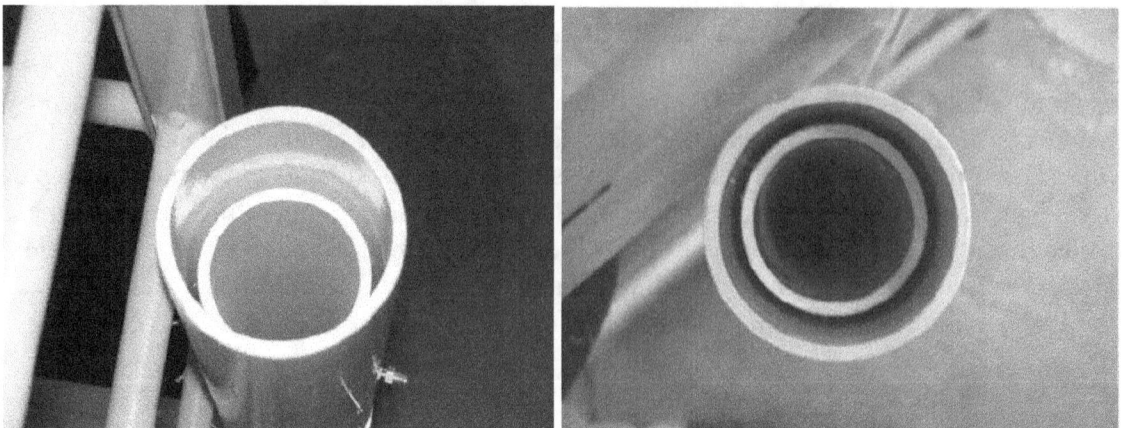

**Figure A.1      Isometric and overhead views of the annular flow path.**

Figure A.2 shows the full length of the test section. A total of seven pressure ports were installed along the length of the outer pipe at intervals of 304.8 mm (12 in.). The bottom port, Port 7, was located 914.4 mm (36 in.) above the end of the outer pipe to ensure all measurements were taken after the flow was fully developed. As an added precaution, all measurements were taken at Port 6 or above.

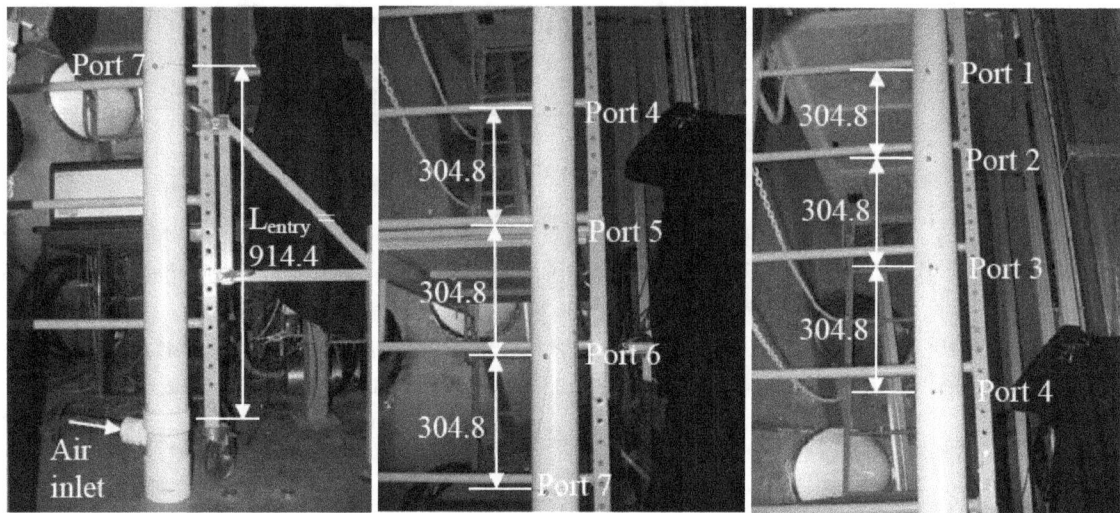

**Figure 2      Photographs of the annular test section. All dimensions are in millimeters.**

Curve fits to the pressure drop data were used to determine the $S_{LAM}$ coefficient of the test assembly. The major, or viscous, pressure loss is expressed in Equation A.1. The length of the flow section, air velocity, and air density are denoted by L, V, and $\rho$, respectively.

$$\Delta P_{major} = f\left(\frac{L}{D_H}\right)\left(\frac{\rho \cdot V^2}{2}\right) \qquad \text{A.1}$$

The friction factor for laminar flow from Kays and Crawford is written explicitly as[5]

$$f = \frac{S_{LAM}}{Re}, \text{ where } S_{LAM} = 96 \text{ (annular flow)} \qquad \text{A.2}$$

Substituting for the Reynolds number yields

$$\Delta P_{major} = S_{LAM}\left(\frac{L}{D_H^2}\right)\left(\frac{V \cdot \mu}{2}\right) \qquad \text{A.3}$$

Curve fits to pressure drop data are presented in the format of Equation A.4.

$$\Delta P_{major} = a_1 \cdot V \qquad \text{A.4}$$

The $S_{LAM}$ coefficient may now be determined explicitly.

$$S_{LAM} = 2 \cdot a_1\left(\frac{D_H^2}{L \cdot \mu}\right) \qquad \text{A.5}$$

The following is an example $S_{LAM}$ analysis of the curve fit to pressure drop data. Please refer to Figure A.2 for the location of the pressure ports described next. The data in Figure A.3 refer to the pressure drops across pressure ports 6–1, 5–2, and 4–3 with lengths of 1524 mm (5 ft.), 914.4 mm (3 ft.), and 304.8 mm (1 ft.), respectively. Using Equation A.5 and the linear curve fit coefficients, the $S_{LAM}$ coefficient was determined to be 95.7, 98.8, and 100.2 for the pressure flow length segments of 6–1, 5–2, and 4–3, respectively. The uncertainty in these coefficients is estimated to be $u_{SLAM} = \pm 3.4$ based on the effects of uncertainties in the mass flow controller and pressure gauges on the curve fit coefficients.

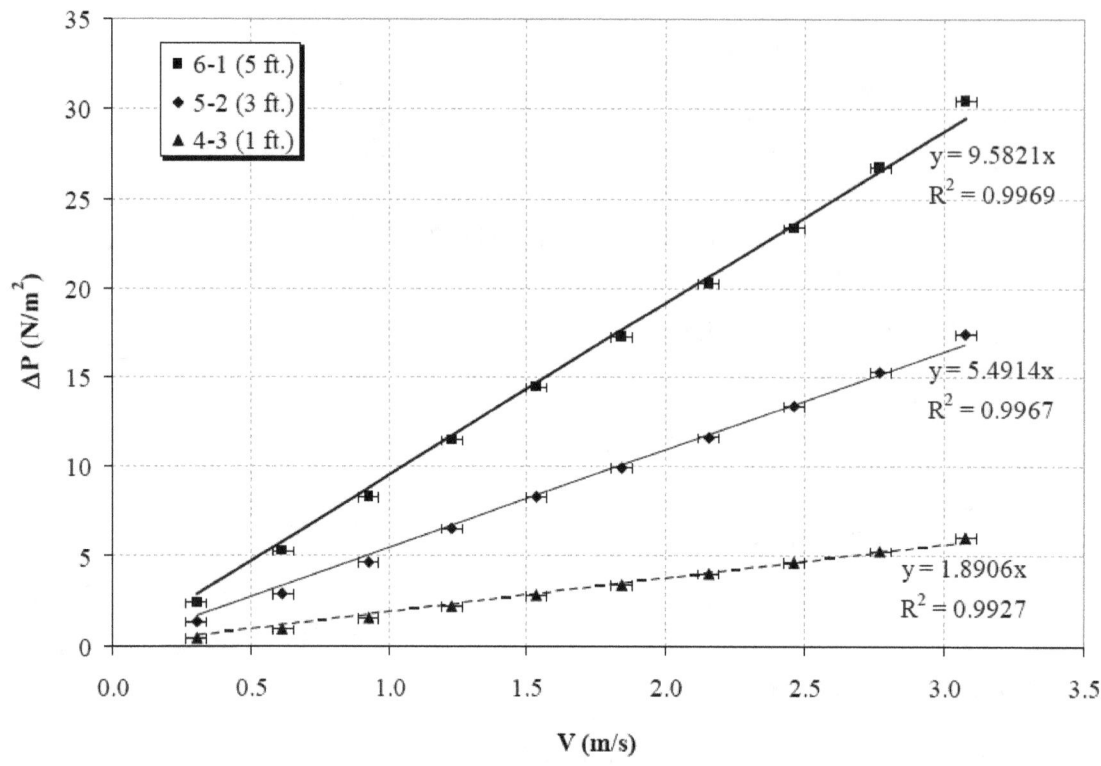

**Figure A.3**      Pressure drop as a function of velocity in the annular region for different flow lengths.

Table A.2 summarizes the data collected for this validation effort. All the $S_{LAM}$'s determined from this analysis matched the analytic value of 96 (Kays and Crawford) within experimental uncertainty with the exception of the first run between ports 6–1.[5] These ports were closest to the inlet and outlet, possibly encountering some secondary pressure losses.

**Table A.2**    $S_{LAM}$ coefficient analysis data for pressure drops between ports 4–3, 5–2, and 6–1 in the annular flow region.

| Pressure Drop | Run | L (mm) | $a_1$ (N·s/m$^3$) | $S_{LAM}$ |
|---|---|---|---|---|
| 4–3 | 1 | 304.8 | 1.8906 | 98.8 |
| 4–3 | 2 | 304.8 | 1.8621 | 97.4 |
| 4–3 | 3 | 304.8 | 1.8487 | 96.7 |
| 5–2 | 1 | 609.6 | 5.4914 | 95.7 |
| 5–2 | 2 | 609.6 | 5.4296 | 94.6 |
| 6–1 | 1 | 1524 | 9.5821 | 100.2 |
| 6–1 | 2 | 1524 | 9.4793 | 99.1 |

Finally, the pressure drop data were plotted against the analytic solution. Figure A.4, Figure A.5, and Figure A.6 show the pressure drop as a function of Reynolds number for the 4–3 (1 ft.), 5–2

66

(3 ft.) and 6–1 (5 ft.) annular lengths, respectively. These data generally conform to the analytic solution with some noticeable deviation above a Reynolds number of approximately 1500. This increased pressure drop could indicate the beginning of the transition to turbulence.

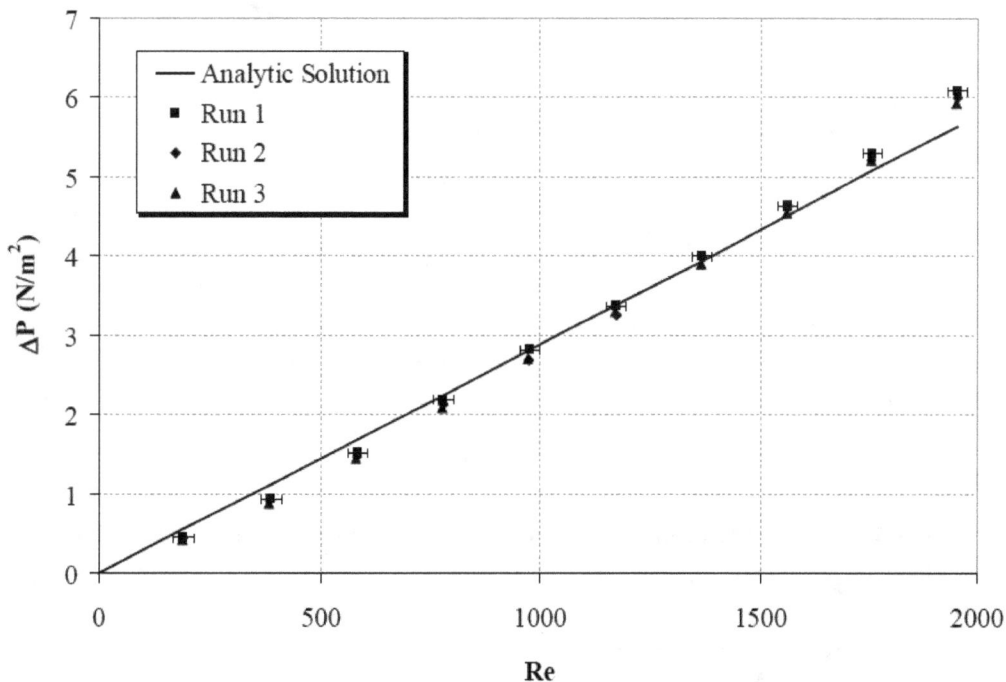

**Figure A.4**     **Pressure drop as a function of Reynolds number across ports 4–3 (1 ft.).**

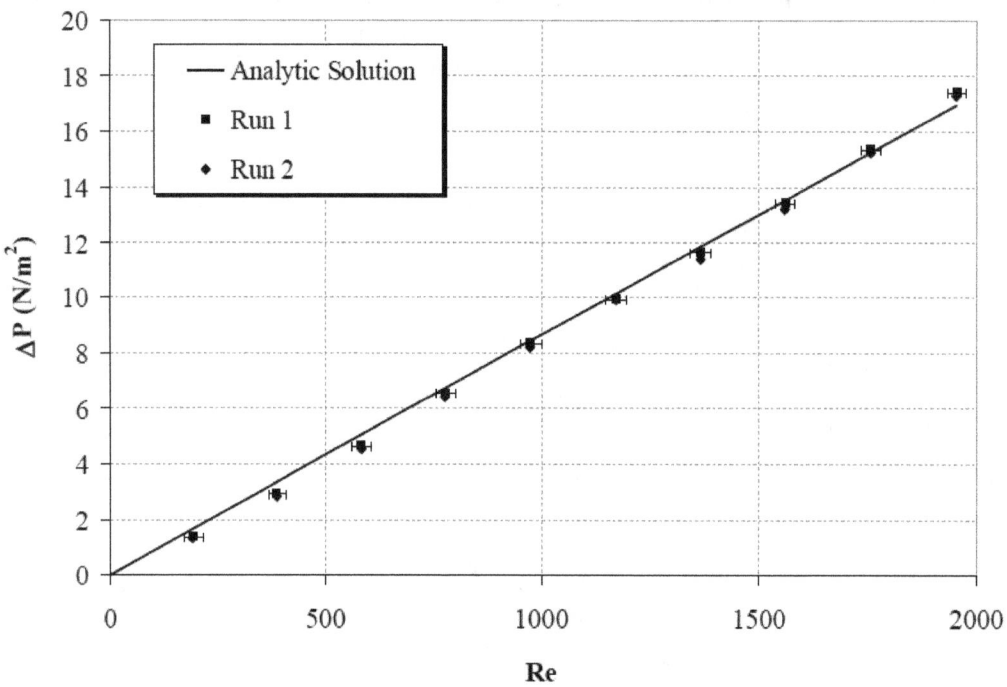

**Figure A.5**       **Pressure drop as a function of Reynolds number across ports 5–2 (3 ft.).**

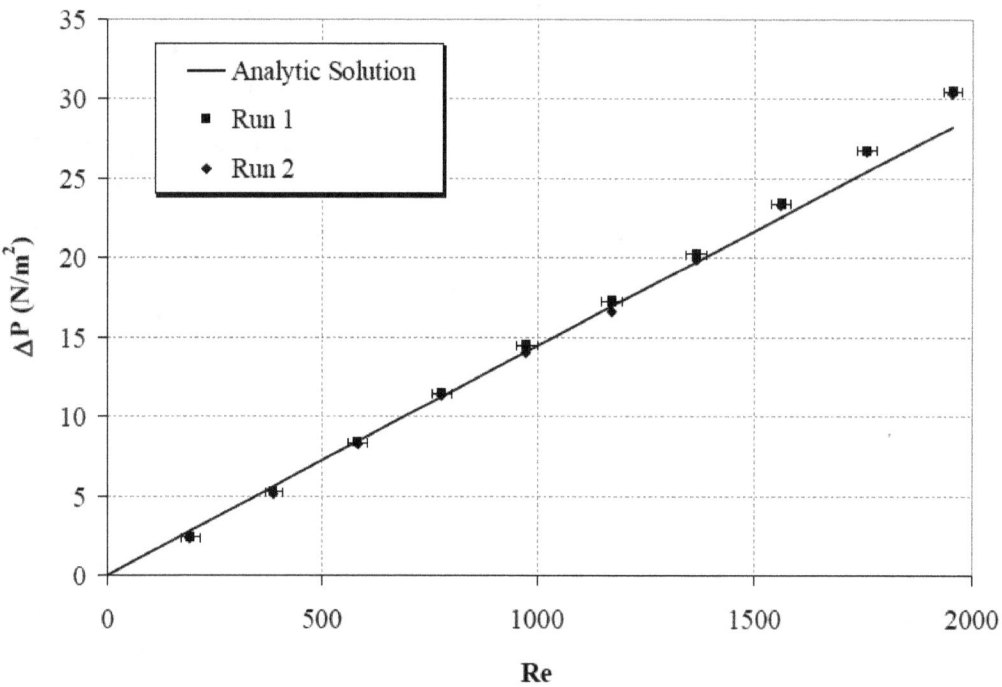

**Figure A.6**       **Pressure drop as a function of Reynolds number across ports 6–1 (5 ft.).**

# APPENDIX B.  ERROR ANALYSIS

The error and uncertainty inherent to an experimental result are critical to the accurate interpretation of the data.  Therefore, the uncertainties in the experimental measurements are estimated in this section.  Results of this analysis are given, followed by a general description of the method used and a brief explanation of the source of each reported measurement uncertainty.

The overall standard uncertainty of an indirect measurement $y$, dependent on $N$ indirect measurements $x_i$, is defined in Equation B.1.  The standard uncertainty associated with an indirect measurement is analogous to the standard deviation of a statistical population.

$$u^2 = \sum_{i=1}^{N} \left( \frac{\partial y}{\partial x_i} u_i \right)^2 \qquad \textbf{B.1}$$

Here, $u$ is used to define the standard uncertainty of a measurement and defines the bounds that include 95% of the possible data.

## B.1 Uncertainty in Assembly Velocity for the Apparatus

The uncertainty in the bundle velocity was determined using error propagation analysis (EPA) for the blocked guide tube measurements in the smallest storage cell (217.5 mm).  The assembly velocity was determined from Equation B.2 in which $Q_i$ is the volumetric flow rate in slpm for each flow controller, $A_{Assembly}$ is the cross sectional area of the assembly, R is gas constant for air, T is the ambient air temperature, and P is the ambient air pressure. The first term in the equation represents the conversion from slpm to kg/s.

$$W_{Assembly} = \left( \frac{0.001 \, \rho \, m^3 / 1 \cdot STP}{60 \, s/min} \right) \frac{\sum_{i=1}^{8} Q_i}{A_{assembly}} \left( \frac{R \cdot T}{P} \right) \qquad \textbf{B.2}$$

Equation B.3 gives the relation between the overall uncertainty of $W_{assembly}$ and the contributions from the measurement uncertainties of Q, T, and P.

$$u^2_{W_{Assembly}} = \left( \frac{0.001 \, \rho \, m^3 / 1 \cdot STP}{60 \, s/min} \right)^2 \left[ \sum_{i=1}^{8} \left( \frac{RT}{P \cdot A_{Assembly}} u_{Q_i} \right)^2 + \left( \frac{Q_{tot} R}{P \cdot A_{Assembly}} u_T \right)^2 + \left( \frac{Q_{tot} RT}{P^2 \cdot A_{Assembly}} u_P \right)^2 + \left( \frac{Q_{tot} RT}{P \cdot A^2_{Assembly}} u_{A_{Assembly}} \right)^2 \right] \qquad \textbf{B.3}$$

Table B.1 summarizes the values used to determine the overall uncertainty of the assembly velocity.  The overall uncertainty in $W_{Assembly}$ was found for the highest volumetric flow rate achieved during testing, $Q_{tot}$ = 2300 slpm, at a typical ambient condition of T = 298 K, P = 83,400 N/m$^2$, and $A_{Assembly}$ = 0.0256 m$^2$. The standard uncertainty was determined to be $u_{W_{Assembly}}$ = 0.012 m/s.  The uncertainty was most affected by the volumetric flow rate (Q) and air temperature (T) contributing 41 and 40% of the overall uncertainty, respectively. The uncertainty in the assembly hydraulic area contributed 14% to the overall uncertainty. The remainder was due to the uncertainty in the atmospheric pressure.

**Table B. 1**    **Measurement uncertainties and intermediate calculations for $V_{bundle}$.**

| Measurement, $x_i$ | Standard Uncertainty, $u_i$ | Influence Coefficient $\dfrac{\partial\left(W_{Assembly}\right)}{\partial x_i}$ | Section Containing Explanation |
|---|---|---|---|
| Volumetric Flow Rate, $Q_1$ | 1.0 slpm | $\dfrac{RT}{P \cdot A_{Assembly}}$ | B.1.1 |
| Volumetric Flow Rate, $Q_2$ | 2.0 slpm | $\dfrac{RT}{P \cdot A_{Assembly}}$ | B.1.1 |
| Volumetric Flow Rate, $Q_3$ | 3.0 slpm | $\dfrac{RT}{P \cdot A_{Assembly}}$ | B.1.1 |
| Volumetric Flow Rate, $Q_4$ | 3.0 slpm | $\dfrac{RT}{P \cdot A_{Assembly}}$ | B.1.1 |
| Volumetric Flow Rate, $Q_5$ | 3.0 slpm | $\dfrac{RT}{P \cdot A_{Assembly}}$ | B.1.1 |
| Volumetric Flow Rate, $Q_6$ | 3.0 slpm | $\dfrac{RT}{P \cdot A_{Assembly}}$ | B.1.1 |
| Volumetric Flow Rate, $Q_7$ | 4.0 slpm | $\dfrac{RT}{P \cdot A_{Assembly}}$ | B.1.1 |
| Volumetric Flow Rate, $Q_8$ | 4.0 slpm | $\dfrac{RT}{P \cdot A_{Assembly}}$ | B.1.1 |
| Ambient Air Temperature, T | 1.1 K | $\dfrac{Q_{tot}R}{P \cdot A_{Assembly}}$ | B.1.2 |
| Ambient Air Pressure, P | 110 Pa | $\dfrac{Q_{tot}RT}{P^2 \cdot A_{Assembly}}$ | B.1.3 |
| Assembly Hydraulic Area, $A_{assembly}$ | $1.1 \times 10^{-4}\ m^2$ | $\dfrac{Q_{tot}RT}{P \cdot A_{Assembly}^2}$ | B.1.4 |

## B.1.1 Uncertainty in Volumetric Flow Rate Q

The volumetric flow rate was controlled with eight MKS volumetric flow controllers operated in parallel (Model # 1559A-24174). The uncertainty of the volumetric flow rate was determined from the stated manufacturer's upper uncertainty of 1% of full scale. The uncertainties in flow rate were 1 slpm for flow controller 1, 2 slpm for flow controller 2, 3 slpm for flow controllers 3 through 6, and 4 slpm for flow controllers 7 and 8. The value shown in Table B.1 represents these standard uncertainties associated with the volumetric flow rate.

### B.1.2 Uncertainty in Ambient Air Temperature

The air temperature was measured with a standard k-type TC. The standard uncertainty for this type of TC is $u_T = 1.1$ K.

### B.1.3 Uncertainty in Ambient Air Pressure

The air pressure was measured with a Setra Systems barometer (Model 276). The uncertainty of the ambient air pressure was taken from the manufacturer's calibration sheet, which indicated an uncertainty in the instrument of ±0.1% of full scale (110,000 Pa). Therefore, the standard uncertainty in the pressure reading is $u_P = 110$ Pa.

### B.1.4 Uncertainty in Assembly Cross Sectional Area

The inner dimension of the storage cell was measured to within ± 0.127 mm (0.005 in). The outer diameter of the simulated fuel rods was measured to within ± 0.0254 mm (0.001 in). This tolerance leads to a maximum uncertainty of $5.5 \times 10^{-5}$ m$^2$ in the hydraulic area.

## B.2 Uncertainty in Pressure Drop Measurements

The manufacturer of the Digiquartz pressure transducers used in these experiments lists a *static error band* of ±0.02% of full scale, or ±4.1 N/m$^2$. This error band includes repeatability, hysteresis, and conformance. Furthermore, these error bands consider the *zero-drift* of the instrument over periods of up to 14 years. Conversations with the manufacturer indicate the experimental procedure followed for these investigations, namely the zero flow measurements to correct any zero drift and the relatively short experimental data collection times ($\sim$ 2 minutes), should place the uncertainty in any pressure data closer to the resolution of the instrument, or 1 part per million of full scale. The largest observed zero drifts in the zero flow measurements were less than 0.92 N/m$^2$, which is smaller than the plotted symbols in this report.

## B.3 Uncertainty in $S_{LAM}$ and $\Sigma k$ Coefficients

The following procedure was adopted to determine the uncertainty in the $S_{LAM}$ and $\Sigma k$ coefficients. Because the greatest experimental uncertainty comes from the bundle velocity, the influence of the velocity on the quadratic curve fits was examined. The overall assembly pressure drops for the blocked guide tubes case were curve fit as a function of $W_{Assembly} \pm u_{W_{Assembly}}$. Figure B.1 shows the resulting curve fits to the pressure drop data across the assembly in the smallest storage cell (217.5 mm). Using these curve fit coefficients, the error associated with the $S_{LAM}$ and and $\Sigma k$ coefficients may now be determined. This procedure was also followed for the pressure drops in the other two storage cells.

71

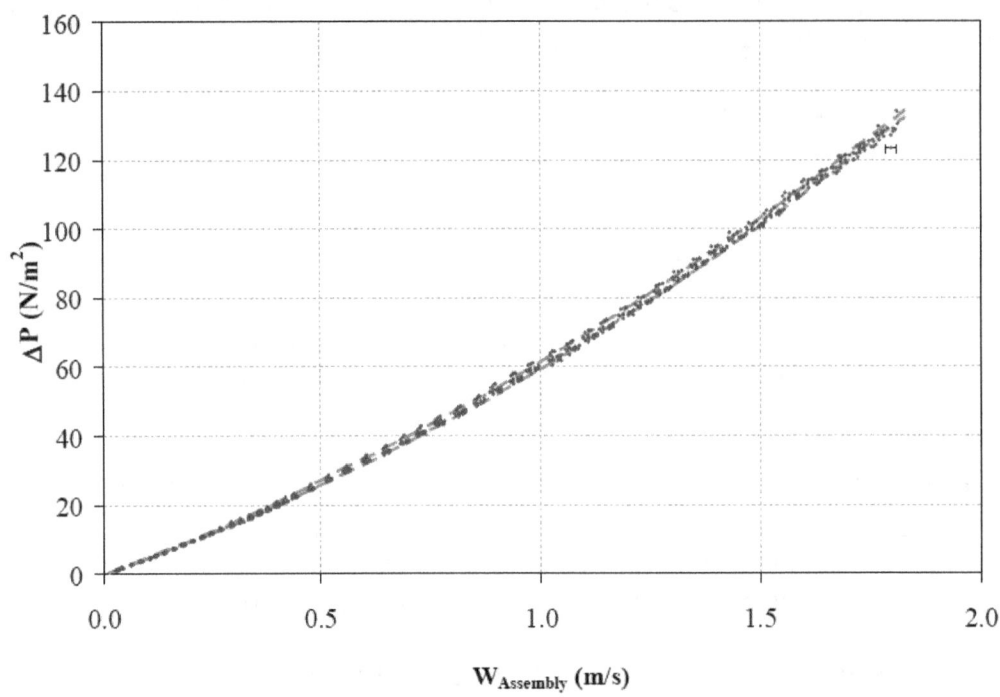

**Figure B.1      Overall assembly pressure drop as a function of velocity for the 217.5 mm storage cell.**

The two dashed curves represent the quadratic fits to the data shifted by $\pm\ u_{W_{Assembly}}$ .

Table B.2 summarizes the $S_{LAM}$ and $\Sigma k$ coefficients determined from this error analysis.

**Table B.2      $S_{LAM}$ and $\Sigma k$ coefficients showing the effect of the uncertainty in assembly velocity.**

| Fitting velocity (m/s) | 217.5 mm | | 221.8 mm | | 226.6 mm | |
|---|---|---|---|---|---|---|
| | $S_{LAM}$ | $\Sigma k$ | $S_{LAM}$ | $\Sigma k$ | $S_{LAM}$ | $\Sigma k$ |
| $W_{Assembly}$ | 133 | 31.2 | 110 | 27.5 | 99 | 27.3 |
| $W_{Assembly}+ u_{W_{Assembly}}$ | 138 | 29.8 | 115 | 26.4 | 104 | 26.3 |
| $W_{Assembly} - u_{W_{Assembly}}$ | 128 | 32.5 | 106 | 28.6 | 94 | 28.3 |

The uncertainty for $S_{LAM}$ and $\Sigma k$ coefficients appears to be slightly dependent on storage cell size. The maximum differences in the $S_{LAM}$ and $\Sigma k$ coefficients in Table B.2 are taken to be the uncertainty, $u_{S_{LAM}} = 5$ and $u_k = 1.4$. These values represent the conservative limit.

| NRC FORM 335<br>(12-2010)<br>NRCMD 3.7 | U.S. NUCLEAR REGULATORY COMMISSION | 1. REPORT NUMBER<br>(Assigned by NRC, Add Vol., Supp., Rev., and Addendum Numbers, if any.) |
|---|---|---|
| | **BIBLIOGRAPHIC DATA SHEET**<br>*(See instructions on the reverse)* | NUREG/CR-7144<br>SAND 2008-3938 |

| 2. TITLE AND SUBTITLE | | 3. DATE REPORT PUBLISHED | |
|---|---|---|---|
| Laminar Hydraulic Analysis of a Commercial Pressurized Water Reactor Fuel Assembly | | MONTH | YEAR |
| | | January | 2013 |
| | | 4. FIN OR GRANT NUMBER | |
| | | Y6758 | |

| 5. AUTHOR(S) | 6. TYPE OF REPORT |
|---|---|
| E.R Lindgren and S.G Durbin | Technical |
| | 7. PERIOD COVERED (Inclusive Dates) |

8. PERFORMING ORGANIZATION - NAME AND ADDRESS (If NRC, provide Division, Office or Region, U. S. Nuclear Regulatory Commission, and mailing address; if contractor, provide name and mailing address.)
Sandia National Laboratory
Albuquerque, NM 87185

9. SPONSORING ORGANIZATION - NAME AND ADDRESS (If NRC, type "Same as above", if contractor, provide NRC Division, Office or Region, U. S. Nuclear Regulatory Commission, and mailing address.)
Division of System Analysis
Office of Nuclear Regulatory Research
U.S. Nuclear Regulatory Commission
Washington, DC 20555-0001

10. SUPPLEMENTARY NOTES

11. ABSTRACT (200 words or less)
To the knowledge of the authors, these studies are the first hydraulic characterizations of a full length, highly prototypic $17\times17$ pressurized water reactor (PWR) fuel assembly in low Reynolds number flows. The advantages of full scale testing of prototypic components are twofold. First, the use of actual hardware and dimensionally accurate geometries eliminates any issues arising from scaling arguments. Second, many of the prototypic components contain intricacies by design that would not be reproduced by using simplified flow elements. While this approach yields results that are inherently specific to the fuel assembly under testing, the differences in commercial designs are considered minor, particularly when considering the hydraulics of the entire assembly.

This report summarizes the findings of the pressure drop experiments conducted using a highly prototypic PWR fuel assembly. The stated purpose of these investigations was to determine hydraulic coefficients, namely SLAM and $\Sigma k$ values, for use in determining the hydraulic resistance in these assemblies within various numerical codes. Additionally, velocity profiles were acquired to estimate the partitioning of flow through the bundle and annular regions within the assembly. The apparatus was tested in the laminar regime with Reynolds numbers spanning from 10 to 1000, based on the average assembly velocity and hydraulic diameter.

| 12. KEY WORDS/DESCRIPTORS (List words or phrases that will assist researchers in locating the report.) | 13. AVAILABILITY STATEMENT |
|---|---|
| Laminar Hydraulic Analysis of a Commercial Pressurized Water Reactor Fuel Assembly | unlimited |
| PWR 17 X 17 | 14 SECURITY CLASSIFICATION |
| SLAM | *(This Page)* |
| Friction Losses | unclassified |
| Viscous Losses | *(This Report)* |
| Inertial Losses | unclassified |
| Flow Losses | 15. NUMBER OF PAGES |
| | 16. PRICE |

NUREG/CR-7144

Laminar Hydraulic Analysis of a Commercial Pressurized
Water Reactor Fuel Assembly

January 2013